CONTENTS

Introduction

Foreword

Having the right skills is essential in all walks of life.

And skills don't come much more important than health and safety, which is why this publication is one of the most important resources CITB produces.

The aim of this revision book is simple – to help construction workers pass their CITB *Health, safety and environment (HS&E) test for managers and professionals*.

Workers who pass the test are highly valued by their employers, colleagues and the industry at large, because it confirms that they have the certified skills to keep themselves – and those around them – safe.

Making sites safe is the responsibility of everyone in construction. Your engagement and leadership are essential to achieving that goal.

Pressures

Construction, by its nature, is a hazardous industry. There's no getting away from that.

While the sector offers fulfilling, well-paid jobs, the pressures of working in our industry can take a heavy toll. Physical and mental illness, and accidents, are all too common.

Figures from the Health and Safety Executive show that construction accounted for nearly a quarter of the 123 work-related fatalities in 2021–2022. Although the number of site fatalities has fallen since last year, more needs to be done to reduce the figures still further.

The mental health statistics for construction are grim, too. The workplace mental health charity Mates in Mind states that over 700 construction workers take their lives each year.

Construction workers and their families should not have to endure this amount of work-related worry, heartache and tragedy.

Responsibility

My CITB colleagues are often asked why managers and professionals must test their knowledge of all areas of health and safety, not just those specific to their role. The reason is clear: construction engineers, architects and managers have a high level of responsibility – and a moral duty – for health and safety.

Even though they do not set people to work on site, and there is often no contractual responsibility for the conduct or management of the site, managers and professionals must know the risks of construction activities, and be equipped with the skills to make their voices heard.

CONTENTS

CONTENTS

Support

CITB is striving to improve health and safety in the construction industry.

Our Business Plan, published in May 2022, shows how we will support health and safety in construction in the future. For example, it gives details about how we will help industry address the challenge of a new building safety regime. We will work to ensure that those involved in design or building work can do their job effectively.

In terms of mental health, CITB is undertaking several initiatives. We are working with the Lighthouse Club to train mental health first-aiders and instructors. We're also partnering with Samaritans to make mental health support accessible to small and medium-sized construction employers, and we're establishing a commission to support apprentices' mental health as they commence their journey into construction.

This work will follow our £90,000 investment, announced in September 2022, to fund a pilot scheme providing mental health support for apprentices from the start of their construction careers.

Benefit

I'd like to thank everyone who has shared their health and safety expertise to develop both the CITB Managers and Professionals HS&E test and this revision book.

I hope you pass the test, and that you put your learning to good use.

The skills you acquire will help you advance in your construction career and, just as importantly, benefit your fellow workers, their families and the public.

Let's keep looking out for each other in the workplace.

Tim Balcon
CITB Chief Executive

The Lighthouse Club can be contacted via email at info@lighthouseclub.org or by phone on 0345 609 1956. Samaritans can be contacted at any time from any phone for free on 116 123.

INTRODUCTION

About this book

This book has been created to help you revise for your Health, safety and environment test. It contains questions that could appear in your test, making it an invaluable resource in helping you pass. It also includes information about how to book your test, any special assistance that is available, and other helpful topics.

To help you get the most from the book, it has been designed for you to use as a workbook. You can test yourself by marking your answers in the book, and then checking these against the correct answers at the back of the book.

About the test

The CITB Health, safety and environment test helps to raise standards across the industry. It ensures that workers at all levels meet a minimum level of health, safety and environmental awareness before going on site.

The test structure has been designed to enable you to demonstrate knowledge across the following key areas.

Section A: Legal and management

Section B: Occupational health, wellbeing and welfare

Section C: General safety

Section D: High risk activities

Section E: Environment

Section F: Specialist activities

Section A: Legal and management

General responsibilities: what you and your employer need to do to ensure that everyone is working safely on site.

Accident prevention: Understand how human, occupational and environmental factors influence accidents.

Construction (Design and Management) Regulations 2015 (CDM): Understand the principal responsibilities under the regulations.

Section B: Occupational Health, wellbeing and welfare

Occupational health and welfare: Understand common construction-related health issues and diseases, and how to mitigate them through the provision of control measures, communication, training, co-ordination and welfare provision.

First aid and emergency procedures: Understand the legal considerations for first aid and emergencies, and what must be made available.

Personal protective equipment: Understand why personal protective equipment (PPE) is always considered the last resort, and the duties and responsibilities.

Dust and fumes (respiratory hazards): Understand how to work safely, protect workers from exposure to respiratory hazards, and the range of health conditions that may arise from exposure to dust and fumes.

Noise and vibration: Understand the risks of excessive noise and vibration in the workplace, and suitable controls.

Hazardous substances: Understand the importance of protecting the workforce from exposure to hazardous substances.

Manual handling: Understand why it is important to handle all materials (loads) correctly, and in compliance with safe systems of work.

Section C: General safety

Safety signage and symbols: Understand the categories of safety signage, as required by the Health and Safety (Signs and Signals) Regulations 1996.

Fire prevention and control: Understand the legal responsibilities for the management of fire risk and prevention on site.

Electrical safety, tools, equipment, lasers and drones: Understand the requirements for and effective use of electrical safety in a construction environment.

Site transport safety and lifting operations: Understand the principles of achieving the safe movement of vehicles, mobile plant, pedestrians and suspended loads on site.

Section D: High risk activities

Working at height: Understand the risks associated with working at height and appropriate mitigation measures.

Excavations and confined spaces: Understand how to work safely in a confined space or excavation.

Section E: Environment

Environmental awareness and waste control: Understand how waste should be managed, and how to conserve energy and resources.

Section F: Specialist activities

You will also be asked questions on the specialist activities of demolition and highway works.

How is the test structured?

All tests last for 45 minutes and have 50 knowledge questions. To pass, you need to answer at least 45 out of 50 questions correctly.

What is a knowledge question?

The knowledge questions cover 17 core areas and two specialist areas that are included in all of the tests. All of the questions are factual, and based on current HS&E areas that impact the construction and built environment sectors. For example, they will ask you to identify risk assessment controls.

You do not need to have an in-depth knowledge of every section of the various statutory regulations that underpin HS&E. However, you do need to evidence that you understand the importance of legislative compliance, what is required of you as a manager or professional, and what you must do (or not do) in certain circumstances - for example, what must be followed in relation to an accident or emergency on site.

Many of the questions refer to the duties of employers. In law, the self-employed can have the same legal responsibilities as employers. To keep the questions as brief as possible, the content only refers to the duties of employers, but the questions apply to both.

Legislation in Northern Ireland and Scotland differs from that in the rest of the UK. For practical reasons, all candidates (including those in Northern Ireland and Scotland) will be tested on questions using legislation relevant to the entirety of the UK only.

There are four different styles of knowledge question that may be presented within your test. These are explained below.

Multiple choice and multiple choice with images

A multiple-choice question will ask you to select one or more answers from a list of options. Some answer options may also contain images.

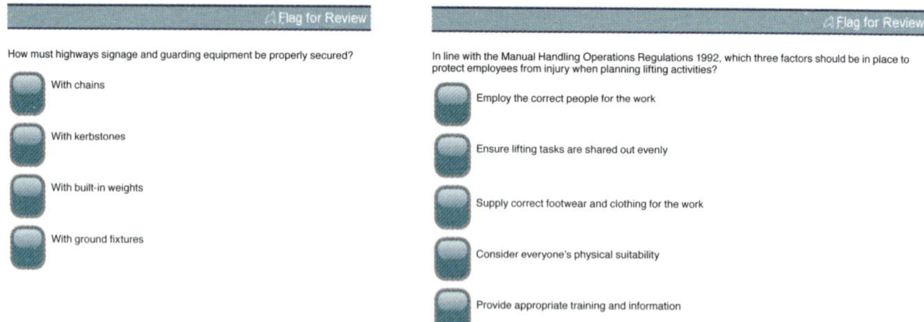

A reasoned judgement question requires you to select one answer from the possible options; these include not only the correct answer, but also the reason for the selection.

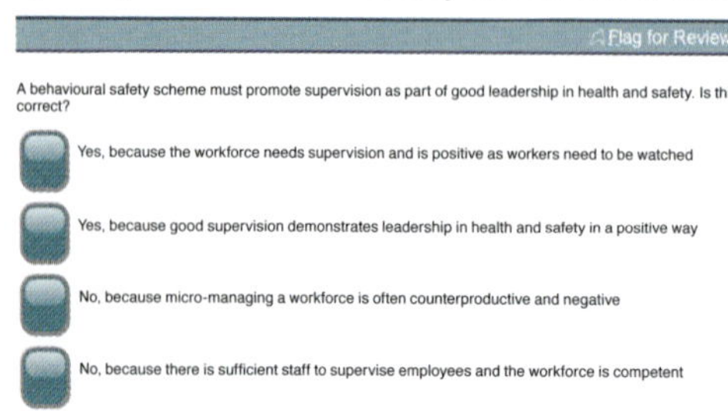

An option selection question requires you to review the options within the question and then select one answer containing the correct combination.

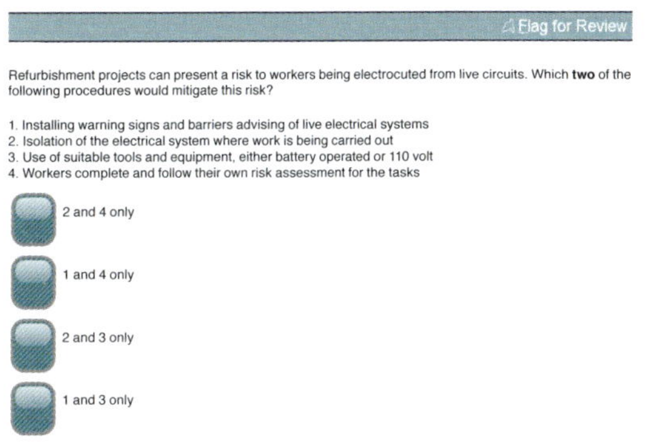
Flag for Review

Refurbishment projects can present a risk to workers being electrocuted from live circuits. Which **two** of the following procedures would mitigate this risk?

1. Installing warning signs and barriers advising of live electrical systems
2. Isolation of the electrical system where work is being carried out
3. Use of suitable tools and equipment, either battery operated or 110 volt
4. Workers complete and follow their own risk assessment for the tasks

2 and 4 only

1 and 4 only

2 and 3 only

1 and 3 only

To practise these question styles online, visit www.citb.co.uk/hsandetest

Who writes the questions?

The question bank is developed by construction health and safety experts alongside assessment and question-writing specialists.

We also work closely with key industry stakeholders to ensure that the content covered in the test is fit for purpose.

Will the questions change?

HS&E legislation, regulations and best practice will change from time to time, but CITB makes every effort to keep the test and the revision material up to date.

- You will not be tested on questions that are deemed to be no longer current
- You will be tested on knowledge questions presented in the most up-to-date edition of the book. To revise effectively for the test you should use the latest edition. You can check which edition of the book you have at www.citb.co.uk/hsandetest or phone 0344 994 4488.

Preparing for a test

To pass your *Health, safety and environment test for managers and professionals,* you need to demonstrate advanced knowledge and a broad understanding of HS&E compliance within a construction working environment. The test itself is broken down into sections, so that knowledge across all key areas can be demonstrated. The revision material is specifically aimed at managers and professionals, and includes topic-related questions that are in the test.

Along with studying this book, there are a number of other ways you can prepare for the test and increase your success.

Revision material	**Managers**
Watch *Setting out*	This video will help you prepare for the behavioural elements that are embedded in the test questions: www.citb.co.uk/hsandetest
Use the digital products	*HS&E test for managers and professionals* - Download (GT200 DL) - App
Read supporting knowledge material	*Construction site safety - The comprehensive guide* (GE700)
Complete an appropriate training course	Site Safety Plus - five-day *Site management safety training scheme*

Where can I buy additional revision material?

CITB has developed a range of revision material, including this book, such as downloads and a smartphone app that will help you to prepare for the test. For further information, and to buy these products:

visit shop.citb.co.uk

call 0344 994 4488

visit retail outlets or online for books. Visit the Apple App store or the Google Play store for smartphone apps

for further products and services that CITB offers, visit citb.co.uk

What's on the app and download?

The app and download offer an interactive package that includes:

- the *Setting out* film
- all of the knowledge questions and answers in both book and practice format
- a test simulator – all of the functionality of the test with the real question bank
- voice overs in English for all questions.

Booking a test

The easiest way to book your test is online. You should be able to book a test at your preferred location within two weeks. You will be given the date and time of your test immediately, and offered the opportunity to buy revision material (for example a book, download or app).

To book your test:

visit www.citb.co.uk/hsandetest

call 0344 994 4488
Welsh booking line 0344 994 4490

When booking your test, it is important that you check the details (including the type of test, the location, the date and time and what ID is required at the test centre), and follow any instructions given regarding the test.

If you do not receive a confirmation email within two hours, please call the booking line to check that your booking has been made.

What information do I need to book a test?

To book a test you should have the following information to hand:

- Which test you need to take
- Whether you require any special assistance (see below)
- Your chosen method of payment (debit or credit card details)
- Your personal details
- Your CITB registration number, if you have taken a Health, safety and environment test before or applied for certain card schemes.

What additional support is available when taking the test?

Language support

The test can be booked with an English voiceover or on-screen Welsh translation. Further details of support available can be found on the CITB website.

How do I cancel or reschedule my test?

To cancel or reschedule your test, you should go online or call the booking number at least 72 hours (three working days) before your test. There will be no charge for cancelling or rescheduling the test online at www.citb.co.uk/hsandetest outside of the 72-hour period. Reschedules and cancellations made via the telephone booking line will incur an administration fee.

Your test fee will not be refunded if you cancel your test less than 72 hours before the test date.

Taking a test

On the day of the test you will need to:

- allow plenty of time to get to the test centre, and arrive at least 15 minutes before the start of the test
- take your confirmation email or letter
- take proof of identity that includes your photo and your signature (such as a driving licence card or passport) – a full list of these requirements can be found on your booking confirmation or online at www.citb.co.uk/hsandetest.

On arrival at the test centre, staff will check your documents to ensure that you are booked onto the correct test. If you do not have all the relevant documents, you will not be able to sit your test and you will lose your fee.

During the test

The tests are all delivered on a computer screen. However, you do not need to be familiar with computers, and the test does not involve any typing. All you need to do is select the relevant answer(s), using either a mouse or by touching the screen.

Before the test begins, you will work through a tutorial. It explains how the test works, and lets you try out the buttons and functions that you will use.

There will also be information displayed on the screen which shows you how far you are through the test, and how much time you have remaining.

After the test

At the end of the test there is an optional survey, which gives you the chance to provide feedback on the test process.

You will be provided with a printed score report after you have left the test room. This will tell you whether you have passed or failed your test, and give feedback on areas where further learning and revision are recommended.

What do I do if I fail?

If you fail your test, your score report will provide you with information on the areas where you got questions wrong.

It is strongly recommended that you revise these areas thoroughly before re-booking. You will have to wait at least 48 hours before you can take the test again.

What do I do if I pass?

Once you have passed your test, you should consider applying to join the relevant card scheme, if you have not done so already. However, please be aware that you may need to complete further training, assessment and/or testing to meet specific entry requirements.

Your score report will also include any areas of the test in which you answered questions incorrectly. It is important that you improve your knowledge in these areas.

To find out more about many of the recognised schemes, visit

www.citb.co.uk/cards-testing/

Fraudulent testing

If you are aware of any fraudulent activity in the delivery of your test, or relating to cards or training in the construction industry email our fraud investigation team at:

report.it@citb.co.uk

CITB takes reports of fraud linked to our testing processes extremely seriously. Working with the Police and other law enforcement agencies, we are doing everything we can to address the issue. Where possible, we always prosecute those engaged in any fraudulent activity.

CONTENTS

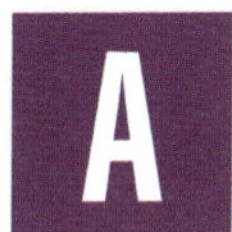

Legal and management

01 General responsibilities

1.01 Which one of the following is a notice that has to be legally displayed on a notifiable building project?

A A notice to stakeholders to erect site boundary hoardings

B A notice of the principal contractor's health and safety policy

C A notice carrying the specified project information

D A notice of health and safety policy statement from the client

1.02 To comply with general site requirements, how long must inspection records be retained for?

A For 3 months after an inspection has been carried out

B For 6 months after an inspection has been carried out

C For 3 months after the project has been completed

D For 6 months after the project has been completed

1.03 You are a contractor on a notifiable project and have not seen the F10 notification form. Which **two** of the following actions must be taken?

A Ensure that the client is aware of their duty to submit an F10

B Ensure the CDM Coordinator has submitted the form F10

C Work within your brief and ignore the lack of F10 notification

D Ensure a form F10 is submitted to the local health authority

E Ensure the client completes and submits form F10 to HSE

1.04 Which of the following **two** duties must a manager ensure is carried out on their site under the Health and Safety at Work etc. Act 1974?

A All workers work safely and independently from each other in their work

B There is a specific number of near misses that are recorded every month

C Any health and safety issues raised on site are reported to the HSE

D Co-operation between all workers on site is ensured for all health and safety matters

E All workers exercise care for themselves and others while carrying out work

1.05 Who is responsible for ensuring that there is suitable welfare facilities on site before any work starts?

A The designer

B The principal designer

C The Health and Safety Executive (HSE)

D The employer

1.06 Which combination of the following make a construction project notifiable to the Health and Safety Executive?

1	Where work will last more than 500 person days	A	1 and 2 only
2	Where 20 or more people will be working on site at any time	B	3 and 4 only
3	Where 30 or more people will be working on site at any time	C	1 and 3 only
4	Where work will last more than 300 person days	D	1 and 4 only

1.07 Which one of the following can lead to prosecution under sections 33(1), (k), (l) or (m) of the Health and Safety at Work etc. Act 1974?

- A Failing to provide a truthful statement to a Health and Safety Executive inspector
- B Failing to update the contractors' all risk insurance policy
- C Failing to engage with workers concerning their health and safety
- D Failing to set up a health and safety committee when requested by two or more persons

1.08 Legislation requires all risk assessments to be?

- A Suitable and detailed
- B Suitable and comprehensive
- C Suitable and sufficient
- D Suitable and technical

1.09 Which **two** of the following can be issued by the Health and Safety Executive or a local authority when a breach of health and safety law has occurred?

- A Improvement notice
- B Compliance notice
- C Letter of authority
- D Prohibition notice
- E Indictment notice

1.10 Identify one of the key duties of the principal designer.

- A To collate construction phase plans, risk assessments and method statements
- B To record the health and safety performance during the construction phase of the project
- C To prepare the health and safety file and revise it when required throughout the project
- D To collate contract documentation and statistics, risk and method statements

1.11 In accordance with Managing for Health and Safety (HSG65), which three elements would be consistent throughout the project life cycle? Select one correct option from below.

A. Element 1: Purpose of bringing a legal proceeding concerning health. Element 2: Safety and welfare. Element 3: Information that may affect national security

B. Element 1: Information relating to workers' occupational health. Element 2: Safety and welfare in terms of good health surveillance. Element 3: Routine check ups

C. Element 1: Leadership and management (including appropriate business processes). Element 2: A trained/skilled workforce. Element 3: Environment where people are trusted and involved

D. Element 1: Information which would be against the interests of national security. Element 2: Matters concerning health, safety and welfare. Element 3: Management on site, including sub-contractors

1.12 Which one of the following lists identify the specific roles that must ensure compliance under the Health and Safety at Work etc. Act 1974?

A. Planning co-coordinator, client and the self employed

B. Employers, the self employed and employees

C. CDM coordinator, employees and employer

D. Principal designer, employer and the client

1.13 If found guilty of an offence, which sections of the Health and Safety at Work etc. Act 1974 could an employer be in breach of?

A. Sections 2 to 6

B. Sections 2 to 12

C. Sections 2 to 8

D. Sections 7 to 8

1.14 General Health and Safety Law emphasises the importance of what is commonly known as the 3 Cs to ensure everyone understands the risks and measures to control risk. Is this correct?

A. Yes, the employer must co-operate, communicate and co-ordinate to ensure effective health and safety is achieved

B. No, the law talks about 5 Cs which include co-operation, co-ordination, communication, concentration and collaboration to meet the health and safety requirements

C. Yes, the employer must co-operate, collaborate and communicate to ensure effective health and safety is achieved.

D. No, the principal contractor must co-operate and co-ordinate effective authorisation of work activities to achieve effective site health and safety

1.15 If there is only one contractor on site, which of the following could be held accountable for breaching the general principles of prevention?

1	The designer	A	1 and 2 only
2	The contractor	B	3 and 4 only
3	The client	C	2 and 3 only
4	The principal contractor	D	1 and 3 only

1.16 Which **two** of the following regulations require a principal contractor to consult with the workforce about health and safety?

A Safety Representatives and Safety Committees Regulations 1977

B Safety Representatives (and Trade Unionists) Regulations 1975

C Health and Safety (Consultation with Employees) Regulations 1996

D Safety Representatives and Safety Consultation Regulations 1978

E Health and Safety Consultation Management Regulations 1993

1.17 If an improvement notice is issued by a Health and Safety Executive inspector, when can work continue? Select **two** answers

A Once the safe system of work has been reviewed, updated and communicated

B When the identified breach has been rectified and re-inspected

C Once additional health and safety controls have been implemented

D Immediately, as long as it relates to ongoing activities; no new work can begin

E Immediately, as long as the work activities are not referred to in the notice

1.18 When must a construction phase plan be drawn up under the Construction (Design and Management) Regulations 2015?

A) When the project is notifiable

B) When the construction phase starts

C) If it is a contractual requirement

D) Before the construction phase begins

1.19 Your company is to carry out refurbishment works on a school that was built in 2011. How would you use the health and safety file as part of the project planning?

A) To confirm the nature, location and markings of significant services, including underground cables

B) To identify hazardous procedures that should be eliminated from the project where possible

C) To identify hazards that have not been eliminated by the original design and construction processes

D) To obtain records and necessary information as to how the original project was designed and built

1.20 A health and safety inspector has arrived at your site. They have witnessed a groundworker using a disc cutter without water-fed dust suppression and have issued them with a prohibition notice.
Was this justified?

A) No – the inspector should have issued an improvement notice, since such work is not an immediate risk to life

B) Yes – the inspector would also give the groundworker a verbal warning and advise them of the risk to their health

C) No – the inspector should have issued an immediate fine along with the prohibition notice, as a regulation has been breached

D) Yes – the inspector would also give the groundworker an affidavit to attend court to defend the breach of the regulation

1.21 Following a fatality on site, a Health and Safety Executive inspector is called to the scene. What would be the first priorities of the inspector?

1	Take samples, measurements and photographs	A	1 and 2 only
2	Ensure everything in the area is left undisturbed	B	2 and 3 only
3	Prosecute the company for health and safety breaches	C	1 and 4 only
4	Review the risk assessments for the work activity	D	2 and 4 only

1.22 On what basis does the Health and Safety Executive charge companies as part of the fees for intervention (FFI) scheme?

A A variable rate for any advice relating to the identified breach

B A daily rate which is linked to the severity of the identified breach

C A fixed hourly rate for the time taken to investigate and identify the breach

D A 'one off' rate for the time taken to investigate and identify the breach

1.23 Site inductions must be reviewed and updated on a regular basis. Why is this important?

A To ensure compliance with policies and procedures

B To align with changing hazards as construction evolves

C To meet the continual needs of quality and compliance

D To ensure engagement and positive outcomes

1.24 Under the Construction (Design and Management) Regulations 2015, the general principles of prevention must be applied by contractors.
Identify **two** ways of how this can be achieved

A Providing dynamic risk assessments for all site tasks

B Providing suitable and sufficient site inductions

C Evaluating the risks that cannot be avoided

D Communicating the construction phase plan to staff

E Adapting to technical advances at every opportunity

1.25 There has been a fatality on site as a result of serious management failings, and the health and safety executive (HSE) prosecution has found direct evidence linking this to an individual site manager. What offence could the individual site manager be found guilty of?

A. Corporate manslaughter and homicide

B. Gross negligence manslaughter

C. Criminal negligence manslaughter

D. Individual negligence manslaughter

1.26 The Health and Safety Executive has made it clear that its enforcement policy is to...

A. actively inspect the businesses that are known to be high risk and have previously paid costs for failing to comply with health and safety law

B. recover the costs for time spent carrying out its regulatory functions from those found to be in material breach of health and safety law

C. target those who cut corners to gain a financial advantage over competitors by failing to comply with health and safety law

D. focus on the financial costs and risks of a custodial sentence for those individuals found guilty of serious breaches of health and safety law

1.27 Current sentencing guidelines help judges and magistrates to decide on appropriate sentences for both health and safety and corporate manslaughter offences. Which one of the following is the focus of the guidelines?

A. Examining culpability, and increasing or decreasing the level of fine according to a range of factors

B. Stronger use of custodial sentences for the individuals found to be guilty of serious breaches

C. Unlimited fines where gross failures in the management of health and safety have caused death

D. Assessment of turnover, in order to set a starting point for a fine to reflect seriousness of offences

1.28 A Health and Safety Executive inspector visiting your site has identified work activities on a scaffold platform with no guardrails or edge protection, and operatives using disc cutters without respiratory protective equipment. Which one of the following enforcement actions is the inspector most likely to take?

A. A formal letter with details of the activity, breaches and action needed to comply with the legislation

B. A prohibition notice to stop the activity either immediately or at the end of the period specified

C. An improvement notice outlining the activity, the work required and the date for completion

D. A prosecution notice informing the employer that it is necessary to prosecute in a magistrates' court

1.29 Under the Construction (Design and Management) Regulations 2015, which document must the principal contractor and contractor keep under review?

A. The health and safety file

B. The designer's risk assessments

C. The project programme

D. The construction phase plan

1.30 In terms of the Plan, Do, Check, Act model, which of the following describes the Plan stage?

A. Profile risks, organise for health and safety, implement your plan

B. Determine your policy and plan for its implementation

C. Plan performance, and monitor and investigate events

D. Review performance, plan, and act on previous lessons learned

1.31 In terms of the Plan, Do, Check, Act model, which of the following describes the Do stage?

A. Determine policy and plan for its implementation

B. Measure and profile performance, and investigate after events

C. Profile risks, organise for health and safety, implement your plan

D. Measure and review performance, and act on lessons learned

1.32 In terms of the Plan, Do, Check, Act model, which one of the following describes the Check stage?

A. To define, communicate acceptable performance and decide what resources are needed

B. To identify controls, assess risks, and record and maintain process safety knowledge

C. To monitor, measure and review performance before events, and investigate after incidents

D. To define, communicate acceptable performance and decide what investment is needed

1.33 Which one of the following is not required to ensure an effective health and safety management system?

A. Relevant organisational processes

B. A competent workforce

C. An environment of trust and engagement

D. An established culture of team work

1.34 The continual development of a company's health and safety policy will help to significantly reduce the common causes of accidents. Is this correct?

A. Yes, as it will identify any shortcomings so appropriate procedures can be implemented

B. No, as it does not provide adequate information for health and safety improvements

C. Yes, as it will detail the day-to-day management for health and safety matters

D. No, as it will only outline the basic elements of health and safety

1.35 An effective health and safety management system includes various elements and practices. Which **two** would not form part of this system?

A. Tender documentation

B. Training and instruction

C. Health and safety file

D. Worker involvement

E. Health surveillance

1.36 Which two of the following health and safety management principles should be given priority?

1	Experience and training	A	1 and 2 only
2	Regular committee meetings	B	3 and 4 only
3	Leadership and management	C	1 and 3 only
4	Supervision and performance	D	1 and 4 only

1.37 Your company employs 30 operatives. Under Part 1 of the Health and Safety at Work etc. Act 1974 (HASAWA), which **two** of the following must be prepared, publicised and revised as often as may be appropriate?

A The company statement outlining its health and safety commitment to employees

B The implementation plan for the structure of the health and safety management system

C The organisation and arrangements for implementing the health and safety policy

D The arrangements for measuring performance and managing health and safety risks

E The statement of intent of the employer's general health and safety policy

1.38 Workers, visitors and members of the public must be protected from welding activities due to the risk of arc eye. Is this correct?

A No, modern welding techniques have eliminated arc eye risk, therefore no risk to others anymore

B Yes, welders must always use welding screens due to the risk of arc eye to protect themselves and others

C No, personal protective equipment (PPE) will remove the risk of arc eye, wearing PPE will protect the welder

D Yes, pedestrian walkways will prevent individuals coming into contact with arc eye and PPE will protect the welder

1.39 Which combination of the following statements are true with regard to permits to work?

1	It must involve two people, the authorised person and the person doing the work	A	1 and 2 only
2	It requires one person and only they can complete the works identified on it	B	3 and 4 only
3	It must be returned to the authorised person for cancellation on completion of the work	C	1 and 3 only
4	It does not need an authorised person to issue it as the work must be completed	D	2 and 3 only

1.40 What is a suitable and sufficient risk assessment? Select **two** options.

- A A simple analysis of any given task
- B A process that eliminates all hazards and risks
- C A method that identifies all foreseeable hazards and risks
- D A way of eliminating dynamic hazards and risks
- E A process that eliminates all unforeseen hazards and risks

1.41 Identify which one of the following contributes to a safe system of work.

- A A method statement
- B A Toolbox talk
- C A permit to work
- D A site induction

1.42 Every employer must make a suitable and sufficient assessment of which one of the following?

- A The number of client and designer representatives based on site, and their exposure to risk
- B The risk of the financial burden placed on shareholders, and the board of directors' dividends
- C The level of workers' safety knowledge and each individual's ability to deliver toolbox talks
- D The health and safety risks that their employees are exposed to while they are at work

1.43 When deciding on the number and location of access and egress points to a site, which one of the following must be considered first?

A Off road parking for cars, vans and visitors

B Providing access for the emergency services

C Access for heavy vehicles, large goods vehicles

D Site security for weekends and bank holidays

1.44 What is the main purpose of a health and safety risk assessment?

A To increase mental wellbeing at work

B To reduce the frequency of inspection by the Health and Safety Executive

C To reduce workplace accidents and ill health

D To increase awareness of accidents and ill health amongst workers

1.45 Which one of the following situations requires a permit to work?

A Entry into a workshop

B Entry into a quarry

C Entry into a confined space

D Entry into a storage compound

1.46 Your company is carrying out emergency out-of-hours response work. Staff will be working by themselves, without close or direct supervision. Which **two** of the following are key considerations when developing safe systems of work for such operations?

A Issuing risk assessments and safety briefings to the workers in a site safety pack

B Giving training on the completion of the dynamic or point of work risk assessment

C Putting appropriate emergency contact arrangements in place in case of an incident

D Giving training on updating generic method statements if any changes are identified

E Amending and updating risk assessments for a specific location or task

1.47 Risk assessments do not necessarily need to be carried out by a health and safety professional. Is this correct?

(A) No. Work tasks can presents complex, unusual or technical issues and unforeseen risks, therefore it is necessary to seek the advice of a competent person

(B) Yes. The amount of effort that needs to be put into carrying out a risk assessment should be appropriate and proportional to the nature of the hazard and level of risk

(C) No. It is essential that all high risks are assessed by a competent health and safety professional as they would be familiar with all aspects of the task being assessed

(D) Yes. When the task being assessed is often repeated or is otherwise familiar, simple or routine, managers and supervisors will often be sufficiently skilled

1.48 According to the Health and Safety Executive (HSE), which one of the following factors relating to health and safety can affect performance and support a management and workforce partnership?

(A) Management and worker partnership based on trust, commitment and positive results

(B) Management engagement with the workforce based on trust, respect and co-operation

(C) Management engagement based on collaboration, cooperation, innovation and respect

(D) Management and worker dialogue and communication to achieve measurable results

1.49 When an organisation wants to achieve successful worker engagement, which two of the following must be embedded?

1	Trust, respect and co-operation between all stakeholders	(A)	1 and 2 only
2	Mutual involvement between management and health and safety representatives	(B)	3 and 4 only
3	Commitment from senior management that is linked to overall performance	(C)	1 and 3 only
4	Motivational reward and loyalty programmes for all workers	(D)	2 and 4 only

1.50 The strategy for managing health and safety in the UK includes genuine management and workforce partnership. Is this true?

A No, because the strategy needs to incorporate the supply chain

B Yes, because without this the strategy will fail

C No, because the strategy also needs to consider unions

D Yes, because strategy is linked to high-performing teams

1.51 A study carried out by the Health and Safety Executive (HSE) found significant errors when reviewing employers post incident risk assessments and highlighted several factors? Which one of the following was identified by this process.

A Too much emphasis placed on reducing personal accidents like slips and trips and not enough on preventing major accidents

B Not enough emphasis placed on minor incidents as these can often have the most significant outcomes in the workplace

C Too much priority to those measures that protect the whole workforce and not focusing on a medium risk level

D Too much priority to low risk and high frequency events as these require significant time and resources to reduce accidents

1.52 Which **two** of the following are responsible for promoting a behavioural safety culture?

A The Health and Safety Executive

B The organisation's shareholders

C The entire organisation

D The organisation's management systems

E The organisation's supply chain

1.53 When organisations have effective and integrated management and leadership development programmes, which one of the following will be the outcome?

A Improved performance management systems

B Access to high-level training courses

C Increased levels of employee engagement

D Enhanced opportunities for career progression

1.54 Which one of the following principles will encourage worker involvement on a construction project?

A Enforcing the need for weekend working

B Challenging and acting on unsafe conditions

C Conducting programme specific toolbox talks

D Reprimanding workers for errors without evidence

1.55 In implementing principles of involvement that have a positive impact on behavioural safety within a workplace, workers must be encouraged to do which one of the following?

A Stop work and be quiet when clients are on the site

B Stop work and report anything that is unsafe on site

C Participate in training courses so they are fully competent

D Use their individual experience to identify unsafe acts

1.56 Implementing a behavioural safety approach is a quick-win initiative that will instantly reduce accidents and increase standards within an organisation. Is this correct?

A Yes - evidence and support for this has come from Health and Safety Executive research into major reportable accidents, which identifies successful outcomes

B No - it takes time and commitment from all levels of a company or organisation, and is not an alternative to applying sound, basic principles of health and safety management

C Yes - companies that incorporate these programmes into the day-to-day running of their businesses see immediate drops in accident rates from the start of implementation

D No - behavioural safety approaches are short-term solutions that change the way organisations work, are time-consuming and stop people from getting on with the job

1.57 To continually improve health and safety on a project, why is effective worker involvement and engagement an important element for success?

1 It evidences commitment and a feeling of value

2 It enhances the organisation's overall reputation

3 It increases the focus on safety over productivity

4 It increases the levels of productivity and quality

A 1 and 2 only

B 2 and 3 only

C 1 and 4 only

D 2 and 4 only

1.58 The rules of health and safety apply not only to employees of a company but also to the...

- (A) general public and local residents
- (B) general public and self-employed people hired by the company
- (C) self-employed people and contractors hired by the company
- (D) local contractors hired by the safety groups

1.59 Which one of the following is more likely to happen if a principal contractor uses contractors and self-employed individuals on a site?

- (A) A lack of adherence and understanding of the health and safety standards and procedures
- (B) A lack of understanding of the worker's health and safety consultation processes
- (C) A lack of clarity and awareness of the emergency and fire contingency procedures
- (D) A lack of compliance with the site's safe systems of work and risk controls

1.60 Identify one of the duties the self employed must follow according to health and safety law?

- (A) They must use their own welfare unit and keep it tidy every day on site
- (B) They must use only their own specific personal protective clothing on site
- (C) They must not cause harm or injury to others when on site
- (D) They must use only battery operated hand tools and equipment on site

1.61 Give **two** examples of how a principal contractor can ensure construction activities are carried out without risks to health and safety?

- (A) Investigate accidents so lessons are learnt
- (B) Hold regular site progress meetings
- (C) Demonstrate good leadership
- (D) Compile all risk assessments
- (E) Audit the supply chain companies

1.62 A behavioural safety scheme must promote supervision as part of good leadership in health and safety. Is this correct?

- (A) No, because micro-managing a workforce is often counterproductive and negative
- (B) Yes, because the workforce needs supervision and is positive as workers need to be watched
- (C) Yes, because good supervision demonstrates leadership in health and safety in a positive way
- (D) No, because there is sufficient staff to supervise employees and the workforce is competent

1.63 Which two of the following must principal contractors put in place to ensure health and safety is continually monitored on a project?

1	Routine inspections due to the changing work environment	A	1 and 3 only
2	Review of approved suppliers	B	1 and 2 only
3	Prompt investigation of near-miss incidents	C	2 and 3 only
4	Revision of site policies on an annual basis	D	1 and 4 only

1.64 Why is it important to include workers in the risk-assessment process for a task that they will be undertaking?

A It will make them more likely to follow any risk-control measures

B It will help them to complete the task within the specified time frames

C It will lead to a significant reduction in task-specific accidents

D It will help them appreciate the importance of following rules

1.65 When selecting suitable contractors for work on a construction project, which of the following will help you find out whether the contractor is complying with their duties under health and safety law?

1	Existing risk assessments from previous similar projects	A	1 and 2 only
2	An independent third-party assessment of their competence	B	1 and 3 only
3	Evidence of any prior accidents and cases of ill health	C	2 and 4 only
4	Membership of a trade association or professional body	D	3 and 4 only

02 Accident prevention

2.01 Which form number must be submitted to the enforcing authority for a reportable occupational disease?

A 2508

B 2518

C 2508A

D 2508B

2.02 Which one of the following occupational diseases does not need to be reported under the Reporting of Injuries, Diseases and Dangerous Occurrences Regulations (RIDDOR) 2013?

A Carpal tunnel syndrome

B Hand-arm vibration syndrome

C Noise-induced hearing loss

D Tendonitis of the hand/forearm

2.03 Under the Reporting of Injuries, Diseases and Dangerous Occurrences Regulations (RIDDOR), which **two** of the following must be actioned if an employee has broken their toe whilst on site?

A The employee must record their injury and details in the site accident book

B The employer must report the injury under RIDDOR within 10 days of the accident occurring

C The employee must inform their supervisor immediately and go to hospital for medical treatment

D The employer must report under RIDDOR if the employee is away from work for over seven days

E The employer must ensure injury-specific toolbox talks are conducted with all sub-contractors

2.04 Which reporting form must be completed by the responsible person after a dangerous occurrence was discovered to be caused by a gasway not being sealed off correctly?

A F2508

B F2508G2

C F2508A

D F2508G1

2.05 Which of the following are the main reasons why construction-related accidents and incidents are recorded in line with the Reporting of Injuries, Diseases and Dangerous Occurrences Regulations (RIDDOR) 2013?

1	To aid the development of risk-control solutions	A	1 and 2 only
2	To analyse the data for budgetary control	B	1 and 3 only
3	To control all associated costs to the business	C	2 and 3 only
4	To enhance the opportunities for repeat business	D	3 and 4 only

2.06 Which **two** types of injuries must be reported to the Health and Safety Executive (HSE) under the Reporting of Injuries, Diseases and Dangerous Occurrences Regulations (RIDDOR) 2013?

- A Broken fingers
- B Specified injuries
- C Over seven-day injuries
- D Over three-day injuries
- E Broken toes

2.07 Is it true that all sub-contractors must notify the main/principal contractor and the relevant enforcing authority of any reportable accidents?

- A Yes, as these are the legal duties of responsible persons
- B No, they only need to inform their immediate supervisor
- C Yes, because accurate records need to be kept by both parties
- D No, as this only needs to be done if the accident resulted in a fatality

2.08 Health surveillance records must be kept for a minimum of...

- A 10 years
- B 20 years
- C 30 years
- D 40 years

2.09 An accident has occurred on site, resulting in injuries to operatives and damage to mobile plant. Under Reporting of Injuries, Diseases, and Dangerous Occurrences Regulations 2013 (RIDDOR) reporting requirements, which of the following incidents require a completed report to be submitted to the appropriate authority, on the approved form, within 10 days?

1	Fires or explosions that cause work to be stopped for 12 hours	A	1 and 2 only
2	Accidents requiring hospital treatment for non-workers	B	1 and 4 only
3	Plant or equipment coming into contact with underground cables	C	2 and 3 only
4	Collapse or failure of load-bearing parts of lifting equipment	D	2 and 4 only

2.10 Which of the following are examples of items that must be reported directly to the appropriate enforcing authority (either the nearest HSE office or the Local Authority), as required by the Reporting of Injuries, Diseases and Dangerous Occurrences Regulations 2013 (RIDDOR)?

1	An employee hits an underground electric cable while operating a road-breaker, causing damage to the sheath of the cable	A	1 and 2 only
2	An employee of a sub-contractor is informed by their doctor that they are suffering from work-related vibration white finger (VWF)	B	2 and 4 only
3	A directly-employed person crushes their hand between two pallets that are being lifted, and breaks three fingers and a thumb	C	1 and 3 only
4	An employee bangs their head and is temporarily knocked unconscious, but is admitted to hospital for 24 hours as a precaution	D	2 and 3 only

2.11 If an incident was to occur on your site, in line with the Reporting of Injuries, Diseases and Dangerous Occurrences Regulations 2013 (RIDDOR), which of the following would you be legally required to report to the Health and Safety Executive?

1	An operative has injuries that will keep them away from work for more than three consecutive days	A	1 and 3 only
2	An operative has injured themselves in a trench, and will spend the next 24 hours in hospital	B	2 and 3 only
3	An operative has injured themselves by crushing both of their hands, and has four fractured fingers	C	2 and 4 only
4	An operative has injuries that will keep them away from work for more than seven consecutive days	D	3 and 4 only

2.12 Latent health and safety failures are caused by the actions of individuals who are not directly involved in operational activities. Is this correct?

A Yes. A latent failure does not impact on the day-to-day health and safety processes

B No. Latent failures involve frontline workers who have breached health and safety some time ago

C Yes. Latent failures occur when management fails to act on a health and safety issue

D No. Latent failures cause significant short and long-term financial costs to the business

2.13 Older workers (those over 50) who have been doing the same job for a long time are often more likely to become injured. Is this correct?

A Yes, as they can become overfamiliar and complacent

B No, as ongoing training would make sure they know the risks

C Yes, as their knowledge becomes outdated and new risks emerge

D No, as they would always act in a safer manner due to experience

2.14 The Health and Safety Executive identifies three main categories of human errors. These are...

A routine, situational and exceptional

B slips, lapses and mistakes

C slips, trips and falls

D lapses, mistakes and violations

2.15 Identify what two things an employer must provide to a new starter once the person has commenced working onsite.

1	Refresher training to make sure new skills are being used on the job	A	1 and 2 only
2	Continuance training to enable them to progress their competence	B	2 and 3 only
3	Team working training to show them how they fit into the team dynamic	C	3 and 4 only
4	NVQ diploma to prove that the employer is an investor in training	D	1 and 3 only

2.16 Why is a construction worker in the older age group demographic who also has considerable site experience at a higher risk of injuring themselves? Select **two** reasons.

- A Slower reaction times
- B Reduced reliance on constant supervision
- C A risk-averse attitude
- D Overfamiliarity with the work
- E Weaker immune system

2.17 How can site managers help to reduce the number of 'situational violations' occurring on a project?

- A Increase routine monitoring
- B Improve the design of jobs
- C Provide more training for emergencies
- D Make procedures relevant and practical

2.18 Young persons (those under 18) are more likely to become injured on a construction site. Is this correct?

- A Yes, they have more of a risk-taking attitude than experienced workers
- B No, they are supervised by a competent person while completing all work
- C Yes, they are not trained to the minimum level to be able to work safely
- D No, they receive the same induction and training as new starters to work safely

2.19 Historically, which type of accident kills the most construction workers?

- A Falling from height
- B Contact with electricity
- C Being run over by site transport
- D Being hit by a falling object

2.20 To ensure that health and safety is not being compromised, how would you show that 'job factors' on a site have been considered at all worker levels and throughout all activities?

1	By implementing safe systems of work	A	1 and 2 only
2	By introducing effective communication methods	B	2 and 3 only
3	By identifying any disabilities or medical conditions	C	1 and 4 only
4	By ensuring adequate time and resources are available	D	3 and 4 only

2.21 Before employing a young person to work on site, which **two** of the following factors should an employer consider when assessing the risks to the young person's health and safety?

A Overfamiliarity with the job or work process involved

B Type of work equipment involved and how it is used

C Slowness of reactions and manual effort required

D Potential for exposure to extreme cold or heat

E General levels of strength, hearing and eyesight

2.22 Which one is not a benefit of accident investigations?

A Understanding how and why things went wrong

B Gaining a realistic snapshot of how work is really done

C Identifying errors in the management of risk control

D Improving internal and external communications

2.23 Who is responsible for recording an accident in the accident book?

A Safety advisor

B Contractor supervisor

C Injured person

D Project manager

2.24 Identify which one of the following is a true statement. The level of accident investigation must be...

- (A) proportionate to the seriousness of the accident
- (B) resourced according to the complexity of the accident
- (C) substantial compared to the size of the accident
- (D) significant compared to the cost of the accident

2.25 Which one of the following must be identified in an accident investigation to ensure it does not happen again?

- (A) The immediate causes
- (B) The underlying causes
- (C) The root causes
- (D) The organisational causes

2.26 Reviewing previous near miss and accident reports can improve safety performance. Is this true?

- (A) Yes, because they identify areas where improvement can and must be made
- (B) No, because acting on everything takes a large amount of time and resource
- (C) Yes, because near misses are often inexpensive to resolve and can impact on safety
- (D) No, because looking at past incidents would lower staff and team morale

2.27 Which one of the following is the least important when investigating a major accident?

- (A) Establishing the full sequence of events
- (B) Inspecting plant for defects
- (C) Identifying types of evidence
- (D) Updating company reporting procedures

2.28 Which of the following is the least important reason for recording all accidents?

- (A) To prevent similar accidents from reoccurring
- (B) To report specified injuries to the Health and Safety Executive (HSE)
- (C) To add them to the accident book
- (D) To prosecute the person responsible

2.29 An incident with the potential to cause harm must be reported to the Health and Safety Executive as a dangerous occurrence, according to the Reporting of Injuries, Diseases and Dangerous Occurrences Regulations 2013 (RIDDOR).
Which one of the following is an example of such an incident?

A Collapse or overturn of load-bearing parts of a forward-tipping dumper

B Mobile scaffold tower coming into contact with overhead power lines

C Explosion or fire that causes work to be stopped for more than 12 hours

D Plant or equipment coming into contact with underground electricity cables

2.30 A completed accident investigation report should be as near as possible to which of the following?

A Comprehensive - including full details of all the data and information obtained

B Concise - based on all witness interviews and the types of evidence identified

C Comprehensive - based on facts and the essential information obtained from witness statements

D Concise - based on facts rather than speculation, unbiased and as accurate as possible

03 Construction (Design and Management) Regulations 2015 (CDM)

3.01 Which one of the following is a client duty under the Construction (Design and Management) Regulations 2015?

A Allow sufficient time and resources for all project stages

B Provide information on remaining project risks

C Estimate the time required to complete work stages

D Review and update the health and safety file

3.02 Under the Construction (Design and Management) Regulations 2015, which one of the following must all duty holders fulfil?

A Co-operation with each other

B Collaboration with their suppliers

C Assessment of project risks

D Planning of their work

3.03 Under the Construction (Design and Management) Regulations 2015, which one of the following is a duty of the principal designer?

A To allow satisfactory time and resources for all project stages

B To make sure welfare arrangements are provided

C To ensure everyone involved in the project is competent

D To estimate the timescales needed to complete work stages

3.04 A principal contractor has successfully won a bid to design and build a shopping centre, and will need more than one sub-contractor working on the project. In line with the Construction (Design and Management) Regulations 2015, which one of the following must the contractors comply with?

A The estimated timescales given to complete the work stages

B The compilation of the construction phase health and safety plan

C The relevant work-related parts of the construction phase plan

D The design, technical and organisational aspects to plan the work

3.05 A contractor has been appointed and they need to obtain pre-construction information from external organisations. Which of the following duty holders can approach these organisations to gather this information?

1 Client

2 Designer

3 Principal designer

4 Principal contractor

A 1 and 2 only

B 2 and 3 only

C 1 and 3 only

D 3 and 4 only

3.06 When a principal contractor is appointed under the Construction (Design and Management) Regulations 2015, which of the following would they be responsible for to ensure effective health and safety?

1	Demonstration of good leadership	A	1 and 2 only
2	Investigation of accidents and sharing lessons learnt	B	2 and 3 only
3	Providing welfare facilities for contractors	C	3 and 4 only
4	Ensuring designers comply with their duties	D	1 and 3 only

3.07 To ensure compliance with the Construction (Design and Management) Regulations 2015, which one of the following dutyholders does not have to comply with the general principles of prevention?

A Clients

B Designers

C Contractors

D Principal designers

3.08 On a construction project involving more than one contractor, what types of information need to be included in the health and safety file?

1	Removal or dismantling of installed plant	A	1 and 2 only
2	Contractual documents	B	1 and 3 only
3	Appropriate surveys and reports	C	2 and 3 only
4	Written systems of work	D	1 and 4 only

3.09 Which one of the following 'general principles of prevention' do **not** need to be applied by principal designers, designers, principal contractors and contractors when conducting their duties in line with the Construction (Design and Management) Regulations 2015?

(A) Providing relevant instructions to employees

(B) Tackling foreseeable risks at source

(C) Giving collective protective measures priority

(D) Co-operating with other duty holders

3.10 Which one of the following actions can a gang of groundworkers take to ensure they are involved in health and safety improvements?

(A) Tell another worker about an unsafe act and decide what to do about it themselves

(B) Identify and report any workplace risks that they come across on site

(C) Make a note of any risks, so they can be mentioned to a supervisor at the next site meeting

(D) Identify workplace risks they come across and speak to the individuals affected

3.11 Under the Construction (Design and Management) Regulations 2015, which one of the following is an additional specific duty placed on principal contractors?

(A) To ensure the adequate supervision of workers

(B) To communicate with all other duty holders

(C) To have worker involvement processes in place

(D) To consult and engage with workers

3.12 A project is due to commence but a principal designer has not yet been appointed. What is the likely negative impact on the project?

- A The residual maintenance requirements may not sufficiently consider safety
- B The allocation of sub-contracted resources may not align to the programme
- C The returned pre-qualification questionnaires may not be adequately examined
- D The vital elements of the health and safety processes may be overlooked

3.13 Under the Construction (Design and Management) Regulations 2015, which **two** of the following will determine the level of supervision, instruction and information required on a project?

- A The behaviours of the workforce
- B The identified project risks
- C The duration of the project
- D The competence of the workforce
- E The number of young people

3.14 A project is at the planning stages, and it has been reported that the client has not yet appointed a principal designer. Why is this significant in regard to the project moving forward?

1	It could have an impact on the successful collaboration between all stakeholders	A	1 and 2 only
2	It could have a negative impact on the gathering of pre-construction information	B	2 and 3 only
3	It could have an impact on the safety of the residual maintenance requirements	C	3 and 4 only
4	It could have an effect on how structured health and safety information is produced	D	1 and 4 only

3.15 On a notifiable project, the co-operation and communication between duty holders is important to ensure health and safety is monitored and controlled.
Is this statement true?

A Yes, otherwise health and safety procedures amongst sub-contractors will become inconsistent and project outcomes will take priority.

B No, senior management would carry out this role and expect all duty holders to co-operate and deliver positive health and safety results.

C Yes, because it will create a positive safety culture where everyone understands the risks and the measures in place to control them.

D No, because all stakeholders involved in the process need to be responsible for their own health and safety arrangements when on site.

3.16 According to the Construction (Design and Management) Regulations 2015, which one of the following is prepared by the principal designer during the pre-construction phase?

A The construction phase plan

B The fire safety plan

C The health and safety file

D The design drawings

3.17 Why is the health and safety plan a significant document during the construction phase of a project?

A It provides a framework for the recording of project information

B It collates all risk-related information

C It specifies control measures where workers could be put at risk

D It controls the effective tendering process

3.18 Which one of the following sources of information would not be included in the Health and Safety File?

A Electrical components drawing

B Steel frame erection method statement

C Air conditioning inspection schedules

D Access procedures to roof voids

3.19 Under the Construction (Design and Management) Regulations 2015, which **two** of the following does an employer, contractor and principal contractor not have to provide workers with?

A Clean and accessible welfare facilities

B Site induction and familiarisation

C A culture of health and safety

D Written outcomes of site meetings

E Access to pre-construction documents

3.20 Under the Construction (Design and Management) Regulations 2015, which duty holder has the responsibility for ensuring the construction phase plan is in place before the work commences?

A Principal contractor

B Contractor

C Client

D Principal designer

3.21 Under the Construction (Design and Management) Regulations 2015, which one of the following does not need to be notified to the relevant enforcing authority?

A The duration of the construction phase

B The planned number of contractors

C The maximum number of welfare facilities

D The peak number of people working on site

3.22 The construction phase plan must detail specific measures for high-risk work activities. Which **three** of the following would fall under this requirement?

A Work in the vicinity of high-voltage power lines

B Work that uses heavy plant and machinery

C Work that poses a risk of drowning to workers

D Work that requires legal health monitoring

E Work that involves energy distribution installations

F Work that involves the removal of demolition waste

3.23 When planning a project it is essential to gather pre-construction documentation. Which **two** duty holders under the Construction (Design and Management) Regulations 2015 have no specific duties in relation to this process?

A Client

B Principal contractor

C Contractor

D Principal designer

E Designer

3.24 When a project is notifiable under the Construction (Design and Management) Regulations 2015, the principal contractor must submit a written F10 notice that specifies project timescales calculated on a five day working week.
Is this correct?

A) Yes, as this is the required process to follow to inform the relevant enforcing authority of a notifiable project

B) No, as the written F10 notice is submitted by the client and the timescales include weekends and bank holidays

C) Yes, as the principal contractor can update the F10 if the project programme runs into weekends and bank holidays

D) No, as the submission of the F10 notice is the responsibility of the client and they determine the working week duration

CONTENTS

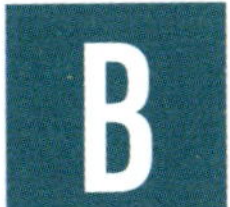

Occupational health, wellbeing and welfare

04 Occupational health and welfare

4.01 You are working in a safety-critical role and have purchased over-the-counter medication due to feeling unwell. What action must you take to protect the safety of yourself and others?

A Inform your supervisor immediately of the medications' possible side affects

B Do not come into work until you have completed the course of medication

C Continue your normal duties, but inform your colleagues of your potential symptoms

D Come in later and leave earlier to reduce the daily hours you are working

4.02 Which one of the following must support a drugs and alcohol policy that is being introduced?

A Education of the harmful effects of misuse

B Arrangements of the 'service level agreement'

C Increased number of mental health first aiders

D Agreement from all supply chain companies

4.03 Why should the control of psychoactive substances be included in a workplace drugs and alcohol policy?

A To ensure the policy is robust and effective

B They fall under the same category of illegal drugs

C They can have the same health effects as illegal drugs

D They are controlled by the Misuse of Drugs Act

4.04 Which one of the following is a reason why recreational drug and alcohol misuse must be considered in a substance misuse policy?

A It is a risk to the individual and others if they are at work and still under the influence

B It is an employer's duty of care to highlight the risks of misuse outside of the workplace

C It will ensure that a prescribed limit is defined to control 'out of hours' misuse

D It will act as a deterrent and reduce the number of recreational users in the company

4.05 Why is it essential to manage the misuse of psychoactive substances in the construction environment?

A To gather data in line with a drugs and alcohol policy to help control health and safety risks

B To mitigate the risks of an individual's altered mental function and emotional state

C To provide evidence of an embedded 'behavioural safety' culture when tendering for future work

D To remain competitive in the sector in order to attract and retain a high level of talent

4.06 If an individual has been under the influence of the drug LSD, which **two** of the following side effects could present a health and safety risk to themselves and others on a construction site?

A Having a flashback weeks, months or years later

B Increasing levels of paranoia and anxiety

C Choking due to a reduced cough reflex

D Becoming temporarily paralysed

E Triggering a mental health condition

4.07 A construction company is introducing a drugs and alcohol (D&A) policy to improve their health and safety arrangements. Which of the following essential components need to be included in this policy?

1	Education and information throughout all levels of the organisation	A	1 and 2 only
2	A summary of the D&A service provider and their collection officers	B	2 and 3 only
3	A thorough training programme for D&A appointed persons	C	1 and 4 only
4	The prevention and rehabilitation support offered to employees	D	4 and 3 only

4.08 When considering the risks of the effects of drug and alcohol use, employers need to consider the implications of misuse outside of the working environment. Is this statement true?

A Yes, employers must make it clear that the misuse outside of the working environment will not be tolerated

B No, employee's conduct outside of the working environment is out of the employer's contractual control as it has no impact

C Yes, recreational misuse by an employee can have an impact on the health and safety of everyone in the working environment

D No, it goes against an employees human rights if an employer interferes and controls recreational misuse

4.09 The legionella bacteria that cause Legionnaires' disease are most likely to be found in which of the following?

A A boiler operating at a temperature of 80°C

B An infrequently used shower hose outlet

C A cold water storage cistern containing water at 10°C

D A cold water supply directly from a wholesome mains

4.10 Which one of the following is the overall purpose of occupational health management?

A Prevention of hazards

B Compliance with regulations

C Well being of workers

D Prevention of physical harm

4.11 Achieving a good level of occupational health on a construction project will be adequately achieved if the...

A health risks are identified at the project design and planning stages

B health risks are discussed at the project planning and execution stages

C external occupational health provider is given an overview of the health risks

D health risk control measures are followed during the execution phase

4.12 To comply with the Control of Lead at Work Regulations 2002, which one of the following will form part of the health surveillance programme?

A) A lung function test

B) A blood test

C) An audiometry test

D) A visual skin examination

4.13 When clearing up debris on a refurbishment project, you fail to notice a discarded hypodermic needle, and it punctures your skin. What must you immediately do to try and mitigate the health risk?

A) Apply suction to the puncture wound for 10 minutes

B) Apply pressure to stop the puncture wound bleeding

C) Encourage the puncture wound to bleed out

D) Tightly dress the puncture wound with a bandage

4.14 When visiting a construction site you observe a positive health culture among the workforce. Which **two** of these practices form part of the organisation's 'health risk management programme'?

A) Occupational health policy

B) Wellbeing awareness campaigns

C) Daily step activity trackers

D) Health surveillance programme

E) Absence management procedure

4.15 Health surveillance and monitoring programmes are the most important elements in regard to managing an organisation's occupational health risks. Is this statement correct?

A) Yes - health surveillance and monitoring are legal requirements, and are the foundation for managing these risks

B) No - although they identify health risks, they are only a part of a much bigger occupational health strategy

C) Yes - health surveillance and monitoring systems are the only accurate way to establish the health risks

D) No - they are not the most important elements, but they need to align to the outcomes of risk assessments

OCCUPATIONAL HEALTH, WELLBEING AND WELFARE

4.16 A construction company is reviewing its minimum standards for occupational health, to ensure that health is managed effectively and that wellbeing is being promoted throughout all levels of the organisation. Which two of the following common principles must be embedded?

1	That the health risks are equal to the safety risks	A	1 and 2 only
2	That any symptoms are managed and controlled	B	2 and 3 only
3	That everyone has ownership and a role to play	C	3 and 4 only
4	That the lifestyle issues are identified and tackled	D	1 and 3 only

4.17 An external consultant has been brought in to assist a large contractor with the introduction of stress management standards. Which two of the following need to be in place before commencing this approach?

1	Commitment and approval from senior management	A	1 and 2 only
2	Establishment of a steering group with membership	B	1 and 3 only
3	Formulation and integration of a behavioural safety policy	C	2 and 3 only
4	Procedures to support effective and open communications	D	4 and 1 only

4.18 Which one of the following could cause occupational dermatitis?

- A Working for long periods of time wearing gloves
- B Contact with another person who has dermatitis
- C Contact with oily clothing and rags
- D Working in the sun without sun protection

4.19 What happens when dry cement is mixed with water?

- A It becomes highly corrosive
- B It becomes extremely toxic
- C It gives off dangerous fumes
- D It becomes a carcinogenic substance

4.20 Which one of the following timbers causes severe dermatitis?

A Boxwood

B Iroko

C Dahoma

D Chestnut

4.21 Why must solvents not be used to clean hands?

A They remove the protective oils from the skin

B They will severely burn the skin

C They will block the pores and ducts of the skin

D They can cause skin cancer

4.22 Identify one cause of irritant contact dermatitis.

A A substance that dries out on the skin resulting in skin damage

B An allergic reaction to something that has come into contact with the skin

C Medications weakening the skin and blood vessels within

D Excessive exposure to ultraviolet light on unprotected skin

4.23 A wellbeing poster campaign entitled 'Working with substances and the importance of clean clothing' is being rolled out on all sites, and aims to raise awareness of the varying health risks linked with poor personal hygiene. Why would working with oil be one of the risks targeted in these posters?

A Oil needs to be washed from the overalls using a harmful detergent

B Oil transferred onto overalls can cause skin problems around the thighs

C Oil can form a barrier on overalls that restricts the effectiveness of cleaning

D Oil transferred onto overalls can trigger ongoing dermatitis problems

4.24 To manage occupational dermatitis on a construction project, which two of the following should be given priority to reduce and control the risks?

1	Provision of protective gloves	A	1 and 2 only
2	Use of equipment for handling substances	B	2 and 3 only
3	Automation of the processes	C	3 and 4 only
4	Regular skin checks for signs of dermatitis	D	4 and 1 only

4.25 Washing your hands thoroughly with soap and water to remove dirt and grime is a preventative measure when working with irritant substances. Is this correct?

A Yes, as regular hand cleaning prevents occupational skin diseases such as dermatitis

B No, washing your hands alone is not enough, you also need to apply a barrier cream

C Yes, as the skins pores, ducts and hair follicles can let irritants enter the sensitive inner skin layer

D No, as the dirt that builds up on the skins pores, ducts and hair follicles gives an added layer of protection

4.26 Which one of the following is a bacterial disease associated with the presence of rats?

A Psittacosis

B Leptospirosis

C Mesothelioma

D Legionella

4.27 Which **two** of the following are symptoms of psittacosis?

A Flu-like symptoms that develop into pneumonia

B Common cold symptoms resulting in paralysis

C Chronic fatigue that develops into muscle spasms

D Severe headache that causes visual impairment

E Acne-like skin condition and ulcers

4.28 You are working near a water course, and you start to feel unwell with flu-like symptoms. Which one of the following actions must you take?

A Go and see your doctor, and explain your working environment

B Continue working, and see if you start to feel better

C Inform the site health and safety advisor

D Take some medication, and have a few days off work

4.29 Identify which one of the following is a risk from roosting pigeons on a construction site.

A If droppings are disturbed, bacteria in their dust and droplets can present significant health risks

B If you disturb roosting pigeons they can become aggressive and are likely to attack

C If the droppings are not removed they can become toxic and present a latent health risk

D They obstruct construction site activities, such as surveying, and cause programme overrun

4.30 When micro-organisms are a potential risk on a construction project, which one of the following models should be used to ensure risks are managed effectively?

A Hierarchy of risk control

B Assess, control and review

C Qualitative risk analysis

D Five-step risk management

4.31 You are working on a brownfield redevelopment site where a number of health hazards have been identified. Which **two** of the following areas on site could pose a risk of workers contracting legionella?

A Areas with a heavy covering of dried pigeon droppings

B Areas where rats are nesting, or have been nesting recently

C Areas with quantities of rust, sludge, scale and organic matter

D Areas with plaster walling that contains infected animal hair

E Areas with redundant water storage tanks, where the water is stagnant

4.32 Individuals who smoke are at a higher risk of contracting Legionnaires disease. Is this statement true?

A) Yes, as smoking severely compromises the immune system making individuals susceptible regardless of their age

B) No, there is no medical evidence that individuals who smoke are at a higher risk from contracting this disease

C) Yes, as smoking damages the lungs, which makes individuals more susceptible to all types of lung infections

D) No, as smoking affects the lungs and immune system, and this disease does not attack the respiratory system

4.33 When testing the water quality of cooling systems, the service provider reported results that identified a risk of legionella growth. Which of the following two conditions in the cooling systems would have identified this risk?

1	The water temperature in some parts of the system was between 20-45 °C	A	1 and 2 only
2	The system was free from deposits of scale and organic matter	B	2 and 3 only
3	The system had evidence of water droplets that could be circulated	C	3 and 4 only
4	The water temperature in all parts of the system was between 30-55 °C	D	1 and 3 only

4.34 Which one of the following organisational factors can be consequences of workplace stress?

A) Inadequate health and safety

B) Improved communication

C) Business efficiency

D) Robust decision making

4.35 You have observed that a member of your team has stopped engaging in meetings and lacks concentration. You are concerned these could be early indicators of work-related stress. As a manager, what should you do next?

A. Discuss your concerns with the employee, make reasonable adjustments and guide them towards an employee assistance programme

B. Discuss your concerns with the employee and ask them to think about how they can improve their engagement and concentration levels

C. Give the employee a copy of the health and wellbeing policy and ask them to choose an intervention that may help them

D. Monitor the situation with the employee for a few more weeks and if the symptoms escalate, have a discussion about support and intervention

4.36 Which one of the following data categories would not be included in a questionnaire aiming to gather data from the workforce on potential stress and anxiety risks?

A. The working environment conditions

B. Design of job roles

C. Reporting of information

D. Mental health first-aid provision

4.37 You notice that a recently-promoted worker has stopped having breaks with the others, and is always last to leave at the end of the day. What could this be an indication of?

A. They could be trying to impress the management on site, because they are looking for a pay rise

B. They could be suffering from stress, and an informal catch-up should be arranged as soon as possible

C. They could be trying to cover up other people's mistakes, and an investigation needs to take place

D. They could be taking materials from the site, and checks need to be carried out immediately

OCCUPATIONAL HEALTH, WELLBEING AND WELFARE

4.38 When developing a questionnaire to gather information from employees about work-related stress, which one of the following question categories is the least likely to gather reliable and valid data?

A The level of communication and sharing of information

B The physical working environment

C The type of work and how it is designed

D The access to training and awareness courses

4.39 Organisational standards that focus on positive ways to reduce work-related stress are being revised and communicated to all employees via a toolbox talk. Which **two** behaviours need to be discouraged during these talks, to ensure that everyone plays their part in tackling workplace stress?

A Deliberately withholding important information

B Turning a blind eye to a potential issue

C Being involved in light-hearted banter

D Overpromising on work targets

E Refusing to attend relevant training courses

4.40 A supervisor has spoken to you about feeling stressed, and has asked you for help with their diagnosis and treatment options. Can you do this?

A Yes - you are responsible for proactively assessing stress levels, as a business priority

B No - the responsibility sits with the individual feeling stressed, who must seek appropriate help

C Yes - you have received extensive training to help with diagnosis and symptom management

D No - you need to escalate this to the external occupational health provider for guidance

4.41 Which one of the following must be in place when providing shower facilities on a site?

A Shower cubicles that can be locked and are sufficient in number for both men and women to use in a communal welfare area

B Shower cubicles that can be used by men and women but separated by a partition within a communal welfare area

C Shower cubicles that can be used by men and women but separated by a partition within a communal welfare area

D Shower cubicles that can be used by men and women at pre-booked time slots which can be used by only one person at a time

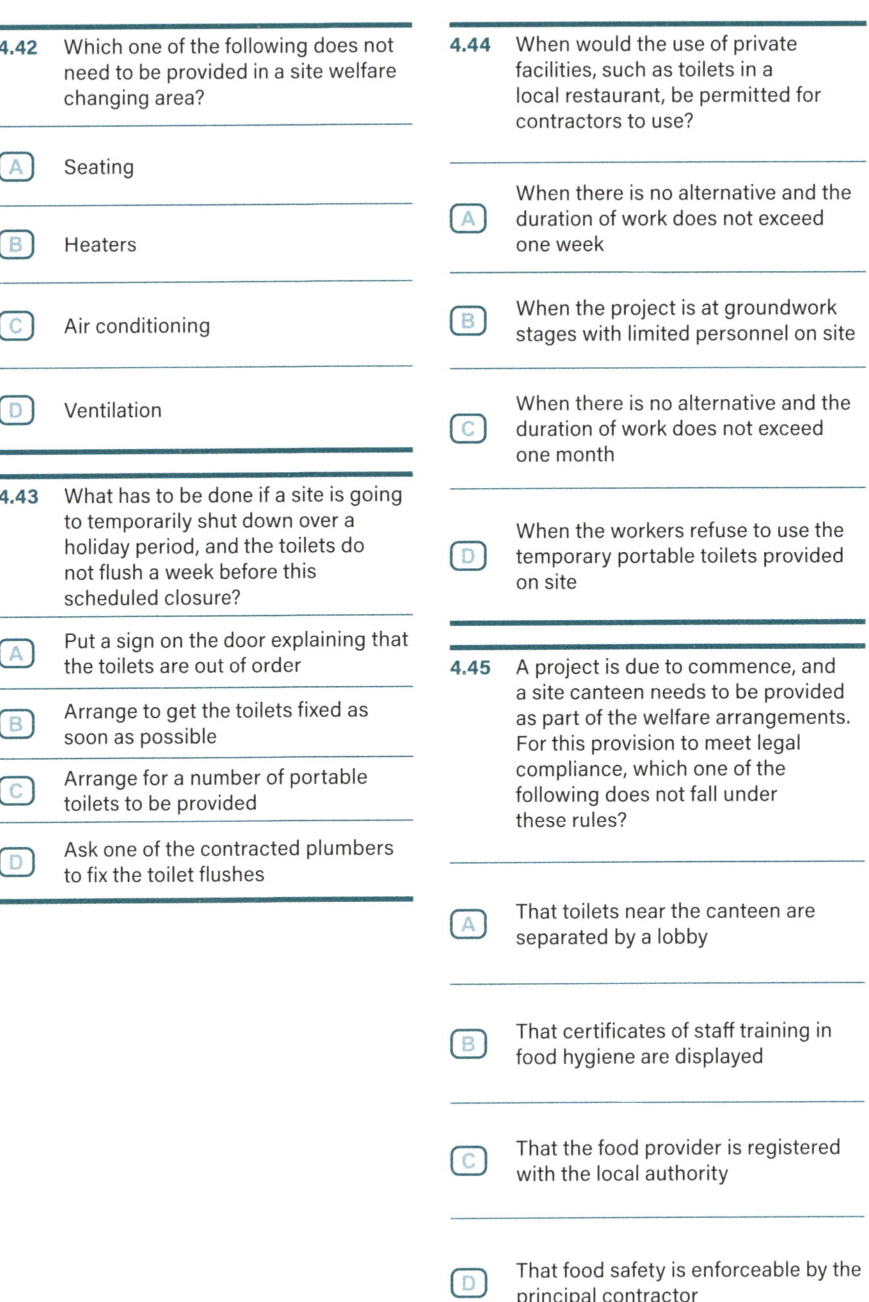

4.42 Which one of the following does not need to be provided in a site welfare changing area?

A Seating

B Heaters

C Air conditioning

D Ventilation

4.43 What has to be done if a site is going to temporarily shut down over a holiday period, and the toilets do not flush a week before this scheduled closure?

A Put a sign on the door explaining that the toilets are out of order

B Arrange to get the toilets fixed as soon as possible

C Arrange for a number of portable toilets to be provided

D Ask one of the contracted plumbers to fix the toilet flushes

4.44 When would the use of private facilities, such as toilets in a local restaurant, be permitted for contractors to use?

A When there is no alternative and the duration of work does not exceed one week

B When the project is at groundwork stages with limited personnel on site

C When there is no alternative and the duration of work does not exceed one month

D When the workers refuse to use the temporary portable toilets provided on site

4.45 A project is due to commence, and a site canteen needs to be provided as part of the welfare arrangements. For this provision to meet legal compliance, which one of the following does not fall under these rules?

A That toilets near the canteen are separated by a lobby

B That certificates of staff training in food hygiene are displayed

C That the food provider is registered with the local authority

D That food safety is enforceable by the principal contractor

4.46 To ensure legal compliance of the welfare arrangements when completing a project site set-up, which **two** of the following must be taken into account?

A That all the facilities are detailed in the construction phase plan

B That the opening hours of the canteen can cater for peak project numbers

C The cleaning and maintenance regimes of the facilities

D The consequences of vandalism highlighted in the health and safety policy

E That a fully stocked first aid room is provided and staffed by a qualified nurse

4.47 The Food Safety and Hygiene (England) Regulations state that food intended for sale or supply must adhere to specific temperature controls to ensure safe levels. Which of the following are the two regulated temperatures?

1	Must be hot, at or above a minimum temperature of 63°C	A	1 and 2 only
2	Must be cold, at or below a maximum temperature of 3°C	B	2 and 3 only
3	Must be warm, at or above a minimum temperature of 37°C	C	1 and 4 only
4	Must be chilled, at or below a maximum temperature of 8°C	D	3 and 4 only

4.48 As part of the site set-up of basic welfare facilities, which of the following must be in place to accommodate men and women?

1	Washbasins must be equal in number to the number of toilet cubicles	A	1 and 2 only
2	Showers must be available in case dirty work occurs during the build	B	2 and 3 only
3	Facilities must be pre-booked via an electronically-auditable rota system	C	3 and 4 only
4	Washbasins must allow for the washing of face, hands and forearms	D	1 and 4 only

4.49 Who has a role to play in promoting fairness, inclusion and respect in the construction sector?

A Clients

B Contractors

C Site managers

D Everyone

4.50 Which one of the following is an example of indirect discrimination in the workplace?

A An employer who does not recruit a visually impaired person as they do not like guide dogs

B An employer who does not interview an applicant due to their race and cultural background

C An employer who only gives full-time employees the chance of promotion and further development

D An employer who unfairly treats an employee over others because they are a family member

4.51 Workplace harassment of one of your colleagues is occurring. Although you are not the target of this harassment, you are legally entitled to pursue a claim.
Is this statement correct?

A Yes, as long as you have the evidence that an offensive environment is being created

B No, it is not involving you so is not detrimental to your working environment

C Yes, as you feel it is not acceptable and it needs to stop immediately

D No, it is just a case of banter between colleagues that has gone to far

4.52 When promoting an equality and diversity culture on site, what must employers provide to enhance health and safety productivity and improved working conditions?

A Clearly-understandable safety information and messages

B A process for workers to follow to suggest improvements

C A series of toolbox talks for staff that focus on the subject

D Giving certain workers responsibility for raising awareness

4.53 Under the Equality Act 2010, which one of the following is the main legal duty for an employee?

A To treat co-workers and contractors consistently and without bias

B To treat managers and co-workers fairly and without prejudice

C To raise any concerns with managers about bullying and harassment

D To treat co-workers, clients and the public with dignity and respect

4.54 How can the 'protective characteristics' under the Equality Act 2010 benefit an organisation? Select **two**.

A They reduce the risk of claims

B They define clear expectations

C They increase staff retention

D They attract highly skilled workers

E They raise awareness about inclusion

4.55 When an employer makes changes to working practices, these improve customer service as well as recruitment and retention, but also...

A comply with legal requirements, to accommodate a diverse workforce

B allow for flexible working opportunities, to retain a competitive edge

C ensure continual improvement in technological advancement

D align and build collaborative relationships with trade unions

4.56 When an employer is raising awareness of fairness, inclusion and respect (FIR) in the workplace, which two legal responsibilities must they comply with in order to protect their employees?

1 They are accountable for harassment of their employees by third parties

2 They must attend annual certified courses on the Equality Act 2010

3 They must act promptly to challenge unacceptable behaviours

4 They should monitor procedures and clearly communicate changes

A 1 and 2 only

B 2 and 3 only

C 1 and 3 only

D 3 and 4 only

05 First aid and emergency procedures

5.01 When displaying notices and signs to inform everyone of the site first-aid arrangements, what needs to be considered?

A They are luminous and can be seen in a variety of weather and light conditions

B They are understandable for workers who have language or reading difficulties

C They are only placed at the site entrance and in the welfare areas

D They contain the number of the local health and safety executive (HSE) officer

5.02 When planning an emergency response to an incident involving a casualty, what do all workers on site need to know?

A How to place the casualty in the recovery position

B The location of first-aid kits and first aiders

C How to submit an online incident report

D The location of the nearest ambulance station

5.03 What would be the most appropriate way to inform staff and visitors about the location of first-aid facilities on site?

A Make sure they read the site notices

B Ask them to speak to the site manager

C Make sure they attend the site induction

D Ask another site worker to tell them

5.04 You have been asked to deliver a site induction to an architect (who will visit regularly) and a group of students (who will visit once). What would be your main consideration?

1	The students will only need to know about the main hazards as they are visiting once, and they will be escorted at all times	A	1 and 2 only
2	Both sets of visitors will need to be escorted at all times; only permanent members of the workforce can have unescorted status	B	3 and 4 only
3	Both sets of visitors will be subject to the same hazards, risks and control measures, which means they can attend the same induction	C	2 and 3 only
4	The architect will need a thorough and detailed induction, as they will be visiting regularly and will likely need unescorted visitor status	D	1 and 4 only

5.05 You are delivering a site induction to new workers, and are trying to highlight the importance of the site's sign-in and sign-out process. What **three** explanations would you give to support this?

A That failure to comply with this requirement may result in disciplinary action being taken against offenders

B That the main purpose of this is to make sure that all persons are accounted for in the event of an emergency

C That it is a legal security requirement under the Counter-Terrorism and Security Act 2015, and that non-compliance could shut the site down

D That this is used as a timekeeping tool, and inaccurate entries could result in a reduction in pay or non-payment of overtime

E That the sign-in and sign-out sheets must be sent to the Health and Safety Executive (HSE) at the end of each month, as part of health surveillance monitoring

F That anyone unaccounted for in an emergency will be treated as missing, and may put emergency workers at risk when looking for someone who isn't there

5.06 Regular fire evacuation drills help to inform workers of crucial emergency procedures. Should the drills happen at the same time each week?

A. Yes - this will accurately measure the efficiency of the procedures in place, and the effectiveness of the evacuation training

B. No - drills are not required every week: it is just as effective to discuss the evacuation process with the workers

C. No - workers can prepare for the drill, which reduces its realism and fails to test workers' knowledge of the evacuation process

D. Yes - this allows workers to pre-empt the evacuation, and shut down or isolate any machinery that they may be using

5.07 Site induction training should cover key health and safety aspects such as first-aid arrangements. Why should training for young employees be a priority?

A. They will not have any experience of first-aid

B. They are particularly vulnerable to accidents

C. They have a poor perception of health and safety

D. They are more likely to take illegal substances

5.08 Your site has activities that include excavations and working at height. How would you comply with legislation and ensure that new workers involved with these activities were aware of the hazards and emergency procedures?

A. By referring to Schedule 5 of The Construction (Design and Management) Regulations 2015, and asking them to read the emergency procedures document

B. By referring to Schedule 2 of The Construction (Design and Management) Regulations 2015, and distributing a leaflet that contains information on the hazards, locations, and emergency procedures

C. By referring to Schedule 1 of The Construction (Design and Management) Regulations 2015, and erecting signs detailing hazards and emergency procedures

D. By referring to Schedule 3 of The Construction (Design and Management) Regulations 2015, and developing a site induction that is specific, highlights risks, and includes emergency procedures

5.09 You have asked a lone worker to carry out a survey in an isolated project location. Which one of the following should you provide them with?

A. An office first-aid kit

B. The location of the nearest hospital

C. A first-aid manual

D. A travelling first-aid kit

OCCUPATIONAL HEALTH, WELLBEING AND WELFARE

5.10 When should an automated external defibrillator (AED) be used on a person?

A When they are suffering from severe shock

B When they are having a stroke

C When they are suffering from heavy bleeding

D When they are having a sudden cardiac arrest

5.11 Where must eyewash stations and burns kits be located?

A Only in the areas that might need them

B In the site office with the site manager

C At the site entrance with the security guard

D Where they can be easily accessed by all workers

5.13 Which **three** factors would determine the need for having an automated external defibrillator (AED) in the workplace?

A How many of the workforce are or have been regular smokers

B How many people may be working in or visiting your workplace

C The location of the site in relation to the nearest medical centre

D Whether any workers suffer from an asthmatic condition

E Whether any work activities could cause amputation of a major limb

F The age, health conditions and medical history of the workforce

5.12 The potential hazard of flying fragments produced by grinding operations has been identified. Mains tap water is not readily available for eye irrigation. What should be included in the site first-aid box?

A A set of sterile, clinical tweezers with rounded magnetic tips and finger stabilising supports

B At least 1 litre of sterile water or sterile normal saline (0.9%) in a sealed, disposable container

C A 500 ml can of compressed, clinically sterile air, in a sealed, water-cooled, reusable container

D At least 1 litre of clinical strength disinfectant or sterile normal saline (2%) in a sealed container

5.14 Is it good practice to include foil blankets in a site's first-aid kit?

A Yes, foil blankets should only be included during the winter months when it is wet and cold, to prevent a casualty developing hypothermia

B No, foil blankets are bulky and take up too much room in a first-aid kit, which means that essential items will need to be left out

C Yes, foil blankets help to retain body heat and they can be used in winter to keep a casualty warm, or in summer if a casualty has gone into shock

D No, if you have developed an effective emergency procedure plan and you have appropriate first-aid facilities, then foil blankets are not needed

5.15 What should be the main consideration when placing first-aid equipment around site?

A It is located in the main site office at all times

B It is located in the first-aid room equipment cabinet

C It is located in a place where it is likely to be needed

D It is located with the first-aider at all times

5.16 Which of the following are key features of an automated external defibrillator (AED)?

1	They are capable of interpreting a casualty's heart rhythm and delivering an electric shock, with minimal input from the operator	A	1 and 2 only
2	They are capable of providing effective cardiopulmonary resuscitation (CPR), which will keep a casualty stable until medical assistance arrives	B	3 and 4 only
3	They are the only device that can prevent death from a heart attack, and have a high success rate for casualty recovery in the UK	C	1 and 4 only
4	They are programmed not to deliver a shock to a casualty unless detecting the presence of a heart rhythm that requires defibrillation	D	2 and 3 only

5.17 Which one of the following must a first aider not do?

A Cardiopulmonary resuscitation (CPR)

B Stop any bleeding

C Administer prescribed medicines

D Treat burns

5.18 Which one of the following is a main responsibility of an appointed person?

A Carry out first-aid duties if a qualified first aider is not present

B Assist the qualified first aider with triage, and monitor all unconscious casualties

C Contact the emergency services and direct them to the scene of the accident

D Apply plasters, dressings and slings to minor wounds and burns

5.20 Which one of the following would be the most suitable to help you identify your first-aid requirements?

A British Red Cross First Aid at Work Manual

B First Aid Approved Code of Practice (ACOP)

C Health and Safety (First Aid) Regulations 1981

D The Workplace (Health, Safety and Welfare) Regulations 1992

5.21 Which **two** of the following factors must be considered when providing first-aid facilities on site?

A The cost of training staff to become qualified first aiders

B The hazards, risks and nature of the work being carried out

C The health, fitness, and medical history of all staff and visitors

D The space in the site office to store the necessary equipment

E The maximum number of people expected to be on site at any time

5.19 A first aider can assist an individual to take their prescribed injectable medicine, but they must never administer it themselves. Are there any exceptions to this rule?

A Yes, a first aider can inject any medicine as it is a skill covered on the first-aid course, but they must gain consent from the casualty first

B No, a first aider would be charged with grievous bodily harm (GBH) if they were to inject any medicine into a casualty

C Yes, a first aider can administer injected adrenaline, such as an EpiPen, for the purpose of saving the life of a casualty

D No, a first aider must never inject any medicine. They should call 999, make the casualty comfortable and wait for professional help

5.22 Although a first-aid certificate is valid for three years, the Health and Safety Executive (HSE) recommends that first aiders undergo annual refresher training. What is the main reason for this?

A To assess whether an employer has a proactive approach to the health, safety and wellbeing of staff

B To provide evidence of continual training, which will help to prevent a prosecution after an incident

C To give employers an opportunity to reduce accidents at work and lower their insurance premiums

D To help qualified people maintain their skills and keep up-to-date with changes to first-aid procedures

5.23 You are conducting a first-aid needs assessment on your site, which has 110 workers. How many first aid at work (FAW) trained staff would you need to comply with The Health and Safety (First-Aid) Regulations 1981?

A Two - one FAW for every 60 workers

B Three - one FAW for every 50 workers

C Four - one FAW for every 30 workers

D Five - one FAW for every 25 workers

5.24 Following a first-aid needs assessment, which of the following additional training is likely to be relevant on all types of construction sites?

1 Application of haemostatic dressings and/or tourniquets

2 Management of a casualty suffering from hydrofluoric acid burns

3 Management of a casualty suffering from cyanide poisoning

4 Management of a casualty suffering from hypothermia or hyperthermia

A 1 and 2 only

B 3 and 4 only

C 2 and 3 only

D 1 and 4 only

5.25 Although the assessment of first-aid does not need to be formally recorded, why is it considered good practice to do so?

A If an accident occurs, employers can use their written documents as evidence during legal proceedings, which will prevent them from receiving a fine or custodial sentence

B The information will help the first aiders on site to choose the contents for the first-aid boxes, and to identify and purchase any specialised medical equipment

C Employers can demonstrate to a safety representative, the Health and Safety Executive (HSE), or local authority inspector, how they decided on their level of first-aid provision

D Employers can print the findings of the first-aid needs assessment and distribute them to visitors or new workers during the site health and safety induction

5.26 What should an employer do to identify the emergency medical provision that is required on site?

A Consult with the nearest hospital

B Conduct a first-aid needs assessment

C Ask a health and safety inspector

D Read a first aid at work manual

5.27 Which **three** of the following factors should be considered in a first-aid needs assessment?

A The skills, knowledge and experience of the workers

B The type of work or operations being carried out

C Whether there are special or unusual hazards

D The minimum number of people that will be on site

E The fitness and agility levels of the workers

F The remoteness of emergency medical services

5.28 You are responsible for providing first aid on site, including a first-aid room. A manager asks you if they can store some electrical equipment in there, as it is rarely used and is a waste of valuable space. Would you agree to this request?

A No - the signals from the electrical equipment could interfere with specialist first-aid kit

B Yes - so long as the equipment was stored neatly in a corner and was always kept clean

C No - the first-aid room should be available at all times, and used only for administering first aid

D Yes - as long as access to the room is not restricted, and equipment is stored neatly under the treatment bed

5.29 Which of the following, other than required first-aid materials, can sites also provide to ensure a satisfactory first-aid room set-up?

1	A sink with hot and cold running water	A	1 and 3 only
2	A resuscitation training mannequin	B	2 and 4 only
3	A reclining chair that offers back support	C	3 and 2 only
4	Drinking water with disposable cups	D	1 and 4 only

5.30 Which one of the following should not be a consideration when selecting first aiders?

A Friendly and reassuring disposition

B Ability to remain calm in an emergency

C Ability to work in an enclosed space

D Acceptable to male and female staff

OCCUPATIONAL HEALTH, WELLBEING AND WELFARE

5.31 First-aid provision must be 'adequate and appropriate in the circumstances'. What should you do to meet this requirement?

A. Provide a qualified first aider for every 25 members of staff, and regularly review the contents of the first-aid needs assessment

B. Instruct a qualified first aider and an appointed person to identify the provision that is needed for the site and specific work areas

C. Conduct a first-aid needs assessment and communicate the contents of the assessment to all site workers and the nearest hospital

D. Ensure that sufficient first-aid equipment, facilities and personnel are available at all times, taking account of alternative working patterns

5.32 When conducting a first-aid needs assessment on site, which one of the following would not need to be given as much consideration regarding the first-aid arrangements?

A. If the build sections are spread out across buildings and floors

B. If members of the public are regularly visiting the site

C. If self-employed workers under the control of an employer are on site

D. If inexperienced workers or employees with disabilities are working on site

5.33 You are observing a first aider conducting an emergency response simulation (mock-up exercise), which involves an unconscious worker lying on the floor next to an electrical cable. What is the first thing that the first aider should do?

A. Locate a wooden pole or stick and move the cable away from the casualty

B. Assess the situation and identify any danger to themselves or others

C. Move the worker away from the cable and out of immediate danger

D. Tell the appointed person to get the automated external defibrillator (AED)

5.34 You are planning the emergency escape and exit routes for a site. Which of the following is one of the areas that needs to be considered?

A. They must lead to a safe, open space, which is free from traffic and has basic medical equipment available such as a fully stocked first-aid kit

B. They must be the quickest and most direct way to the assembly area. Torches must be provided at the start, and various other stages of the route

C. They must be kept clear and free from obstruction and, where necessary, provided with emergency lighting so that they can be used at any time

D. They must only be put in place if there are ten or more workers present at any time, and workers will be encouraged to plan their own routes

5.35 Emergency procedures need to be developed for the effective rescue of a casualty if they fall into a net system. Which one of the following do you consider to be a reasonably practicable approach?

A. Any equipment that is required needs to be available within five minutes of an incident. The rescue must only be carried out by qualified first aiders who have rope access training.

B. Any suitable equipment needed has to be on site, but it does not need to be located near the working area. A first-aid kit will be available in the working area and qualified first aiders will know the risks.

C. Any equipment that is required needs to be instantly available in the work area, and suitable. The rescue must be pre-planned and take into account the work activities and the changing site environment.

D. Any equipment in the work area must be able to deal with any type of rescue. All workers will need to be qualified first aiders, and have rope access training.

5.36 A worker has fallen into a safety net and has sustained injuries. Which one of the following procedures should now be used to prevent further injury?

A. Use a mobile access tower (MAT) to gain height access and send a first aider onto the net to assess the casualty

B. Send a first aider and appointed person onto the net to then assess the casualty and apply necessary first aid

C. Pull the netting so it is taut, then send a first aider onto the net to assess the casualty, and wait for the paramedics

D. Use a mobile elevated working platform (MEWP) to gain height access and then bring the casualty down safely

5.37 When using a safety net system, what is the main consideration that should be given to first aid?

(A) There is a first-aid kit located on the edge of the netting

(B) The casualty can receive first-aid treatment while in the net

(C) All operatives working near to the net are first-aid trained

(D) First aid will not be required as the net prevents injury

5.38 Under the Health and Safety (First Aid) Regulations 1981, which of the following are identified as duties of an appointed person?

1	Providing emergency cover, within their competence, where a first aider is absent due to unforeseen circumstances	(A)	1 and 2 only
2	Investigating all first-aid related incidents and producing a report that provides detailed statistics	(B)	1 and 4 only
3	Supporting the first aider to prepare, distribute and administer medication on sites with more than 100 workers	(C)	2 and 3 only
4	Looking after the first-aid equipment and facilities and calling the emergency services when required	(D)	3 and 4 only

5.39 Your risk assessment has identified there will be volatile COSHH hazards on site. Will this impact your first-aid provision?

A Yes - additional first-aid kits will need to be available, and all first aiders will need to undertake advanced trauma management training

B No - the provision should be the same across all sites. Suitable site inductions and toolbox talks will help to raise awareness of the hazards

C Yes - the provision will need to be suitable, and the first aiders should undertake specialised training courses related to the specific hazards

D No - the first aiders will receive burns training on their courses, which will mean that they can effectively manage any incidents

5.40 For a new construction project, which **three** of the following need to be shared with the local emergency services so they are fully informed?

A Precise location of the site

B Predicted ground conditions

C Specialist equipment required

D Location of site entrances

E Copies of rescue procedures

F Specific hazards on site

06 Personal protective equipment

6.01 Which one of the following is not a requirement for impact-resistant safety goggles?

A. Tinting to reduce glare from the task

B. Regular cleaning and maintenance

C. Suitability for the task and individual

D. Markings to identify their type and suitability

6.02 To avoid the risk of flying fragments, what type of eye protection must a groundworker wear when breaking concrete with a pneumatic drill?

A. Impact-resistant goggles

B. Welding helmet

C. Full visor

D. Safety glasses

6.03 A visitor has arrived on site to assess a worker carrying out a grinding activity. They have no eye protection with them. What action must you take?

A. Provide them with impact-resistant goggles

B. Escort them off the site immediately

C. Issue them with standard safety glasses

D. Advise them to return when they have suitable eyewear

6.04 Which one of the following has been identified as the main cause of eye injuries to construction workers?

A. Not wearing the eye protection

B. Not being given suitable eye protection

C. Not storing the eye protection correctly

D. Defacing the eye protection

6.05 When issuing personal protective equipment (PPE) to a worker, which one of the following must be taken into account?

A. That it fits correctly allowing for safe working

B. That it can be modified for maximum comfort

C. That it meets personal choice and preferences

D. That it is easily trackable for cost allocation purposes

6.06 When purchasing eye protectors and shields from a new supplier, it is important they are all individually marked to identify their 'type and suitability'. Is this correct?

A. No, it is the responsibility of the supplier to ensure records are retained of the batch type and suitability

B. Yes, it is one of many requirements to ensure compliance with British and European specifications

C. No, as the process of marking the polycarbonate material could seriously reduce the level of protection

D. Yes, it is a requirement of the British Safety Industry Federation (BSIF) which represents manufacturers

6.07 It is important for an operative to wear suitable eye protection while working with irritant materials to ensure...

A an effective seal around the eye area

B clear visibility at all times

C the eyes do not become dry

D compliance with mandatory requirements

6.08 What are the **two** main benefits to the user when wearing suitable eye protection?

A It will fit them better and provide a better level of comfort

B It will encourage them to look after it and store it correctly

C It will not obstruct movement when working

D It will provide them with clearer and improved vision

E It will eliminate the risk of an eye injury in any situation

6.09 When operating an angle grinder, what is the significance of wearing a BS EN 166 rated face shield, as opposed to safety glasses?

1	It gives protection against medium-energy impacts	A	1 and 2 only
2	It is suitable for withstanding high-speed particles	B	2 and 3 only
3	It allows the air to flow and prevent fogging	C	1 and 4 only
4	It is face-fit tested to give a secure seal	D	1 and 3 only

6.10 What type of safety footwear must be worn when spreading and levelling wet concrete to form a slab floor?

A Steel-toe and mid-sole boots

B Long safety wellington boots

C Waterproof safety trainers

D Safety rigger boots

6.11 When employers are selecting the most suitable protective footwear for specific types of on-site activity, which one of the following must be carried out?

A An assessment of its suitability in line with the nature of risk

B A hazard identification assessment of the work environment

C A cost-projection analysis of its return on investment

D A plan to identify a systematic and timely training programme

6.12 What is the minimum requirement of safety footwear on a construction site?

- A Steel toecaps and steel midsole
- B Laces and steel toecaps
- C Ability to withstand all weather conditions
- D Provision of adequate ankle support

6.13 Which one of the following is a benefit of safety boots with integrated ankle support?

- A Reduction of foot injuries
- B Increased level of comfort
- C Higher level of durability
- D Prevention of foot movement

6.14 Which one of the following is the main reason for steel toecapped safety footwear?

- A To protect the feet from dropped objects
- B To protect the feet from oil contamination
- C To provide protection on uneven surfaces
- D To provide extra cushioning around the toes

6.15 On your construction site, you have to select and issue protective footwear to workers for general operations. Which of the following are key factors for the wearer to consider?

1 Footwear should be slip-resistant with anti-slip soles
2 Footwear should be flexible and water-resistant
3 Footwear should be able to absorb perspiration
4 Footwear should be heat-resistant and insulated

- A 1 and 2 only
- B 2 and 3 only
- C 1 and 4 only
- D 2 and 4 only

6.16 There are many types of safety footwear available for workers on site, but which **two** of the following actions can make a significant contribution towards reducing injuries?

A Providing footwear with steel midsoles and toecaps, to prevent penetration by sharp or dropped objects

B Providing safety footwear that benefits the worker by preventing injuries when walking on loose surfaces

C Providing information, instruction and training to make sure that safety footwear is used properly, and to explain why it is required

D Providing workers with manufacturers' advice on the most suitable footwear for specific types of hazards likely to be present on site

E Providing manufacturers' information safety data sheets on the care and maintenance of safety footwear

6.17 Your company is carrying out refurbishment work that involves oxyacetylene welding and cutting of structural steel. Which **two** of the following are key considerations concerning protective footwear for operatives undertaking this particular type of work?

A Thermal insulation

B Insulating soles

C Non-slip footwear

D Chemical resistance

E Quick-release fastenings

6.18 When selecting protective gloves, what needs to be carefully considered?

A The material, as some are easily penetrated by chemicals

B The number of workers, to ensure sufficient provisions

C The costs from the supplier against budget allocation

D The compatibility with various barrier creams

6.19 Which one of the following would protect the hands and forearms from the high temperatures and sparks created from welding?

A Elbow-length gauntlets

B Cuff-length rubber gloves

C Long-sleeve Kevlar gloves

D Anti-vibration gloves

6.20 You are planning a road reinstatement project, which involves hot works and the use of a bitumen boiler. Which of the following would be the recommended type of glove to use for these specific workplace hazards?

A Asbestos substitute or Nomex gloves

B Long-sleeved Kevlar gloves or gauntlets

C Neoprene or chrome-leather gloves with reinforced palms

D Heavyweight rubber, nitrile or PVC gloves

6.21 You are planning work where operatives have to break up reinforced concrete slabs using handheld pneumatic concrete-breaking drills. For this type of activity, wearing anti-vibration gloves is recommended to achieve the necessary level of protection. Is this correct?

A Yes - only correctly-fitted gloves, approved by the machine's manufacturer, should be used

B No - they reduce the flexibility of the hands, resulting in higher grip and push forces

C Yes - they keep hands insulated when operating machines that cause vibrations, such as drills

D No - to achieve that level of protection will require specially-designed protective equipment

6.22 A risk assessment has identified that hand and arm protection are required for a task involving pressed steel lintels with sharp edges. Which one of the following actions should also be taken?

A Consider if any additional personal protective equipment is required that has not been identified in the assessment

B Comply with the assessment, because it will cover all hazards and conditions on site associated with the task

C Carry out spot checks on workers as they complete the task, to make sure they are following the assessment to the detail

D Implement control measures as identified in the assessment, and consider the accident and emergency procedures

6.23 When selecting protective gloves for a range of specific processes and operations on site, which **two** of the following are examples of mechanical hazards that must be considered?

A Masonry blocks

B Angle grinders

C Serrated blades

D Thin metal sheets

E Petrol disc cutters

6.24 You are carrying out a risk assessment for a task using stain removers to clean paint and graffiti from concrete and brickwork. The work will be conducted in a city centre at night and during the winter. Which **two** types of gloves would be the most suitable for this task?

A Nitrile

B Chrome leather

C Nomex

D Heavyweight rubber

E Reinforced leather

6.25 Which one of the following is the recommended type of personal protective equipment (PPE) to be worn that can assist in preventing head injuries?

A A safety helmet that is kite marked and has a date of production

B A safety helmet that has a date of production and has a chinstrap

C A safety helmet that is compatible with ear defenders and eye protection

D A safety helmet that is comfortable with an interchangeable headband

6.26 When selecting head protection, the Personal Protective Equipment at Work Regulations (PPE) 2022, state that it must be 'suitable'. Which **two** of the following define the term 'suitable'?

A Must be adjustable to fit the wearer and be compatible with other equipment worn at the same time

B Must be of a type that can be stored correctly after its use and maintained in good condition and repair

C Must be appropriate for the risks involved without increasing the overall risk and conditions of exposure

D Must be stamped with its expiry date with a reversible cradle so it can be worn back to front if required

E Must be supported with the user manual and recommendations for its safe maintenance and storage

6.27 The weather forecast has indicated that high winds are likely to hit your area. This could affect workers who are wearing personal protective equipment (PPE), specifically those wearing hard hats. You should now check the original risk assessment (RA) and review the suitability of the PPE, because...

1	The PPE is now a hazard, and the RA needs to be updated to allow for non-use of hard hats	A	1 and 2 only
2	The conditions could affect the performance of the PPE, and additional control measures may be required	B	3 and 4 only
3	The RA would definitely have included high winds as a hazard, and identified the appropriate control measures	C	1 and 4 only
4	The PPE must be appropriate for the risks involved, as well as the conditions it is being exposed to	D	2 and 4 only

6.28 It is a requirement to wear suitable head protection on all building and construction sites, unless...

A You are working inside a finished building which has been made ready for handover

B You are working on a task that involves leaning over exposed edges

C You are working on a task where wearing a safety helmet reduces your visibility

D You are working in an area where there is no risk of injury from falling objects

6.29 When working on a construction site, which one of the following is a benefit of wearing suitable head protection?

A It can increase the probability of a claim and compensation after a head injury

B It can prevent any injuries where the work involves leaning over exposed edges

C It can allow other compatible personal protective equipment to be worn

D It can protect the worker from direct sunlight and reduce heatstroke

6.30 You are delivering a site induction to new workers, and explaining how to check a safety helmet for signs of damage. One of the workers asks whether they can personalise their helmet. How would you respond to this?

A No - this would be classed as interfering with personal protective equipment, and would result in instant dismissal

B Yes - you can write your name on the peak, and your blood group and date of birth on the inside of the helmet

C No - the solvents in some paints, adhesives and indelible markers can reduce the strength of the plastic

D Yes - you can write your name, the date of your site induction and any additional duties, such as first aider

6.31 You are working at height, which involves leaning over exposed edges. What action must you take to ensure that your head protection remains in place?

A Adjust the plastic cradle inside the safety helmet to ensure a tight fit, so that it remains secure

B Reverse the safety helmet, to prevent the wind from getting under the peak and lifting it off

C Make sure that the safety helmet is fitted with a chinstrap, and is being properly worn at all times

D Ensure the safety helmet is fitted with an additional lanyard, so that if it falls it will not hit another person

6.32 On site, you see an operative's safety helmet fall off and drop onto a concrete slab from a height of approximately two metres. Which of the following statements will apply?

A Safety helmets are manufactured and designed to offer a pre-determined level of impact resistance if they fall off, or have been dropped onto a hard surface

B A safety helmet that has fallen from height onto a hard surface may have suffered damage that will affect its strength - even though no cracks are visible, it should be replaced

C A safety helmet that has fallen from height should be inspected by a competent person, and should be replaced if it has been damaged or is showing visible signs of wear and tear

D Many safety helmets have built-in features that enable them to fall or be dropped onto a hard surface without suffering damage that will affect their strength

6.33 Which one of the following must an employer do if noise levels in a working area are at or above the lower exposure action value?

A Ensure all employees in the area wear hearing protectors

B Ensure hearing protectors are available to affected employees

C Ensure employees not involved in the task stay out of the area

D Ensure no employee enters the area until the noise levels are reduced

6.34 When employees are likely to be exposed to noise levels at or above the upper exposure action value (EAV), which one of the following must an employer do?

A Provide awareness on hearing loss symptoms

B Provide ear plugs and encourage their use

C Provide a selection of types of hearing protection

D Provide hearing protection and ensure it is used

6.35 The selection of hearing protection should initially consider which of the following factors?

A Type and performance of the hearing protection

B Type and level of comfort of the hearing protection

C Type and attenuation levels of the hearing protection

D Type and amplitude levels of the hearing protection

6.36 Which one of the following must be considered when selecting suitable hearing protection?

A Suitability for use with mobile phone earphones

B Handling and fitting of the hearing protection

C Information, instruction and training for employers

D Compatibility with other safety equipment

6.37 All workers should be provided with information, instruction and training on how to fit and use disposable ear plugs, but should they be shown how to check the quality of the ear plug fitting?

A Yes - they should know how to conduct a sound check, which requires the use of specialist audiometry equipment

B No - a sound check is not required, as once the worker is exposed to a noisy area, they will know if the fit is good

C Yes - they should know how to conduct a visual depth check, as correctly-fitted plugs will not be visible from the front

D No - they do not need to carry out a visual depth check, as the supervisor will do this for them and advise if the fit is good

6.38 Workers on site are issued with foam earplugs as personal protective equipment (PPE). Which **two** of the following are considerations for management, in order to ensure that the earplugs perform as expected?

A Training in how to fit and wear them

B Provision of a range of different sizes

C Provision of individually-moulded earplugs

D Supervision to ensure that they are worn properly

E Checking that there is no damage to the seal

6.39 Your risk assessment has identified the need for hearing protection on site. If you choose ear defenders for the workers, which factors must you consider?

1	Ear defenders might not be suitable, and may cause ear irritation or infections as they can become excessively dirty	A	1 and 2 only
2	Thick frames on safety eyeglasses can affect the fit and seal of ear defenders, and reduce attenuation levels	B	2 and 3 only
3	Fitting is critical for ear defenders, as they offer an improved level of hearing protection and comfort compared to other earplugs	C	1 and 3 only
4	The practice of wearing earphones under ear defenders may mask auditory warnings and reduce awareness	D	2 and 4 only

6.40 You are assessing the risks of emergency overnight road repairs, where groundworkers need to work near to a large excavator breaking out a concrete slab. The provision of hearing protection in this circumstance would be considered as a last resort. Is this correct?

A Yes - providing personal hearing protection or other types of hearing protective equipment should be considered as a last resort in any circumstances

B No - providing personal hearing protection due to noise should be one of the first considerations on discovering a risk to the health of your employees

C Yes - providing personal hearing protection should only be considered as an alternative to controlling noise by technical and organisational means

D No - providing personal hearing protection should only be a consideration when there is a need for longer-term noise reduction measures

6.41 Class 2 high-visibility clothing is typically worn in which one of the following groups of working environments?

A Airports, highways maintenance and railways

B Sites with vehicle or plant movements

C Roads with a speed of 40 mph or above

D Dual carriageways and motorways

6.42 Work is about to start on a housing development project located on private land. There will be lots of vehicle and plant movements, and access will be needed for delivery drivers. How would you achieve the minimum visibility level required for these hazards?

A Provide high-visibility waistcoats with one 5 cm band of reflective tape around the body

B Provide high-visibility vests with three-quarter length sleeves and two 5 cm bands of reflective tape around the body

C Provide high-visibility vests with two 5 cm bands of reflective tape around the body

D Provide high-visibility long-sleeved jackets with two 5 cm reflective bands around the body and braces over the shoulders

6.43 Which one of the following is the minimum acceptable visibility level for Class 1 high visibility clothing?

A 0.10 m^2 of reflective material

B 0.15 m^2 of reflective material

C 0.17 m^2 of reflective material

D 0.20 m^2 of reflective material

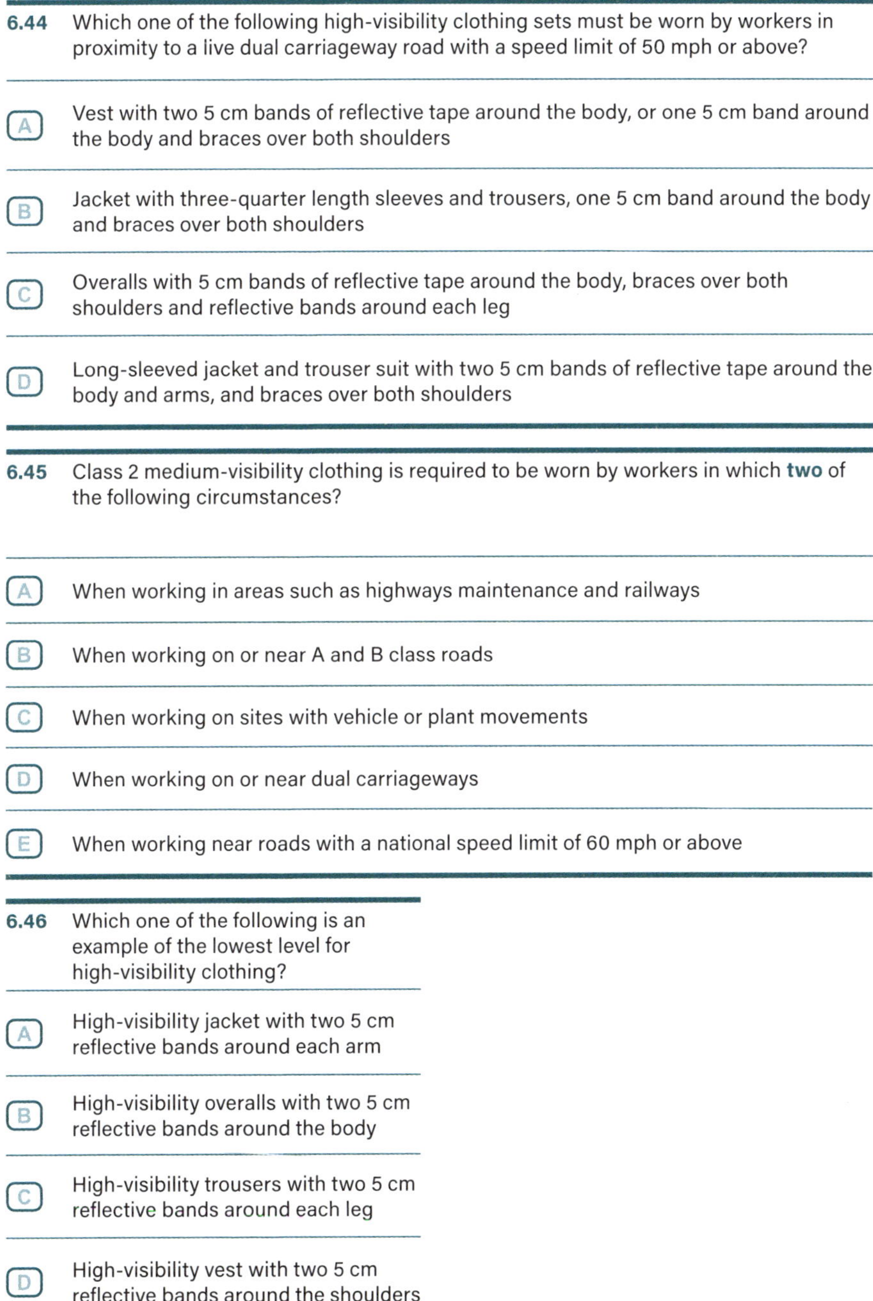

6.44 Which one of the following high-visibility clothing sets must be worn by workers in proximity to a live dual carriageway road with a speed limit of 50 mph or above?

A Vest with two 5 cm bands of reflective tape around the body, or one 5 cm band around the body and braces over both shoulders

B Jacket with three-quarter length sleeves and trousers, one 5 cm band around the body and braces over both shoulders

C Overalls with 5 cm bands of reflective tape around the body, braces over both shoulders and reflective bands around each leg

D Long-sleeved jacket and trouser suit with two 5 cm bands of reflective tape around the body and arms, and braces over both shoulders

6.45 Class 2 medium-visibility clothing is required to be worn by workers in which **two** of the following circumstances?

A When working in areas such as highways maintenance and railways

B When working on or near A and B class roads

C When working on sites with vehicle or plant movements

D When working on or near dual carriageways

E When working near roads with a national speed limit of 60 mph or above

6.46 Which one of the following is an example of the lowest level for high-visibility clothing?

A High-visibility jacket with two 5 cm reflective bands around each arm

B High-visibility overalls with two 5 cm reflective bands around the body

C High-visibility trousers with two 5 cm reflective bands around each leg

D High-visibility vest with two 5 cm reflective bands around the shoulders

6.47 High-visibility clothing is designed to protect the wearer from all harm by making them easily visible to vehicle and plant operators, as well as the travelling public.
Is this correct?

(A) Yes - it is designed to protect the wearer from all physical injuries while working on a site, by visually alerting vehicle drivers and public traffic to the presence of workers

(B) No - it is designed for workers in some specific roles, such as highways operations, and allows their supervisor to check on them easily from a distance in any lighting conditions

(C) Yes - it is designed for workers in industries where a particular background colour is specified for clothing, such as fluorescent orange for railway workers

(D) No - it is designed to make the wearer easy to see under any light conditions in the day, and under illumination such as vehicle headlights in the hours of darkness

6.48 Your employees are going to be carrying out repair works to an airport runway and side-access roads. What significant features must the personal protective equipment issued to your workers provide, when equipping them for this type of work?

1	The clothing provided must be weather appropriate, and include a hood that can accommodate a safety helmet	(A)	1 and 2 only
2	The clothing must be a different high-visibility colour to that issued to airport air-side employees, to avoid confusion	(B)	2 and 4 only
3	The clothing issued must be of the highest visibility level, and meet the Class 3 standard	(C)	1 and 3 only
4	The clothing issued must have a clearly-defined contractor identification band on the back	(D)	3 and 4 only

6.49 Your project involves frequent vehicle and plant movements, and requires working on or near a busy single carriageway A-class road, with a speed limit of 40mph. Which one of the following would be the minimum level of visibility required for this work?

(A) Long-sleeved jacket and trouser suit, with one 5 cm band of reflective tape around the body and arms, and braces over both shoulders

(B) High-visibility trousers, with two 5 cm reflective bands around each leg, and a vest with one 5 cm band around the body

(C) Waistcoat and trouser suit, with two 5 cm bands of reflective tape around the body and arms, and braces over both shoulders

(D) High-visibility vest, with two 5 cm bands of reflective tape around the body, or one 5 cm band around the body and braces over both shoulders

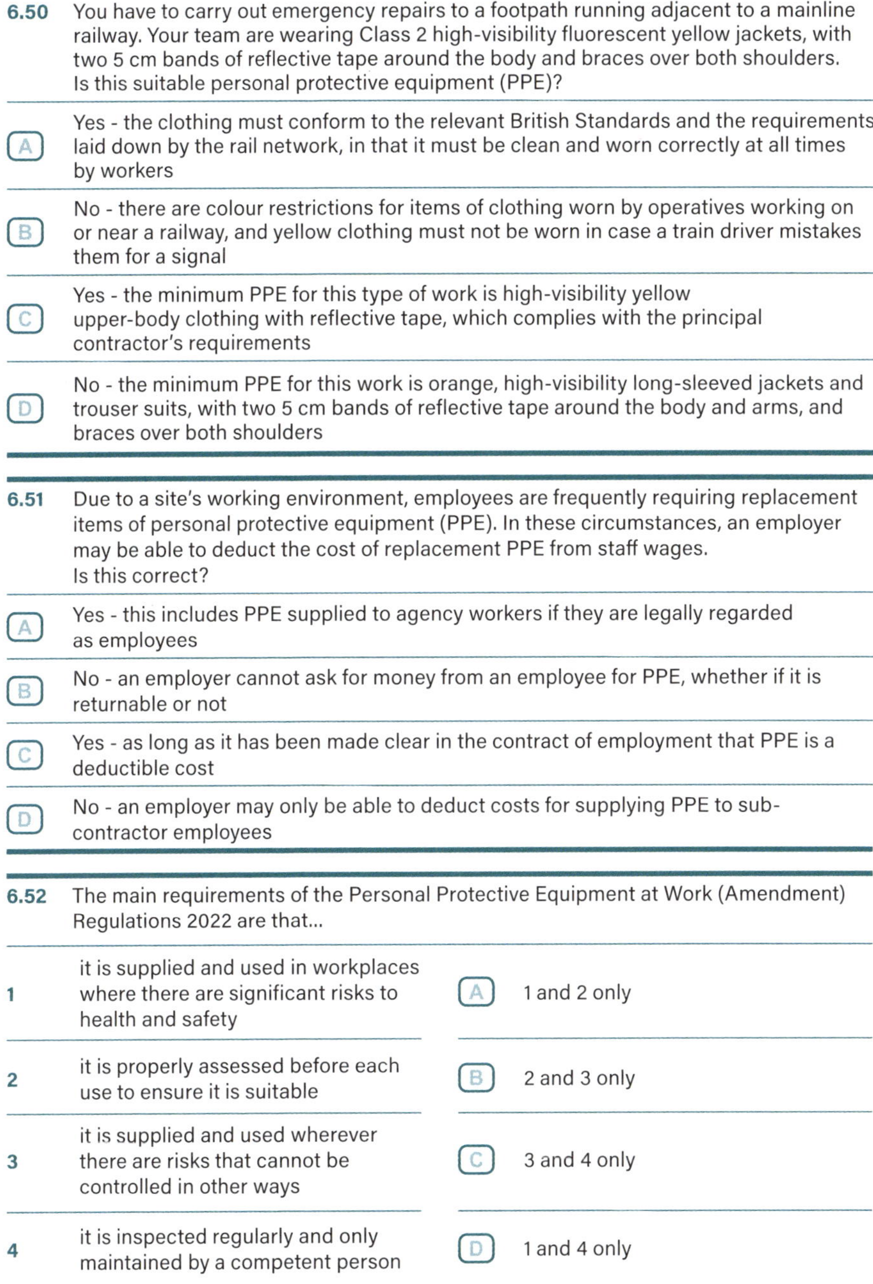

6.50 You have to carry out emergency repairs to a footpath running adjacent to a mainline railway. Your team are wearing Class 2 high-visibility fluorescent yellow jackets, with two 5 cm bands of reflective tape around the body and braces over both shoulders. Is this suitable personal protective equipment (PPE)?

A) Yes - the clothing must conform to the relevant British Standards and the requirements laid down by the rail network, in that it must be clean and worn correctly at all times by workers

B) No - there are colour restrictions for items of clothing worn by operatives working on or near a railway, and yellow clothing must not be worn in case a train driver mistakes them for a signal

C) Yes - the minimum PPE for this type of work is high-visibility yellow upper-body clothing with reflective tape, which complies with the principal contractor's requirements

D) No - the minimum PPE for this work is orange, high-visibility long-sleeved jackets and trouser suits, with two 5 cm bands of reflective tape around the body and arms, and braces over both shoulders

6.51 Due to a site's working environment, employees are frequently requiring replacement items of personal protective equipment (PPE). In these circumstances, an employer may be able to deduct the cost of replacement PPE from staff wages. Is this correct?

A) Yes - this includes PPE supplied to agency workers if they are legally regarded as employees

B) No - an employer cannot ask for money from an employee for PPE, whether if it is returnable or not

C) Yes - as long as it has been made clear in the contract of employment that PPE is a deductible cost

D) No - an employer may only be able to deduct costs for supplying PPE to sub-contractor employees

6.52 The main requirements of the Personal Protective Equipment at Work (Amendment) Regulations 2022 are that...

1	it is supplied and used in workplaces where there are significant risks to health and safety	A	1 and 2 only
2	it is properly assessed before each use to ensure it is suitable	B	2 and 3 only
3	it is supplied and used wherever there are risks that cannot be controlled in other ways	C	3 and 4 only
4	it is inspected regularly and only maintained by a competent person	D	1 and 4 only

6.53 In which of the following **two** circumstances would the Personal Protective Equipment at Work (Amendment) Regulations 2022 not be the first regulations to consider?

A For gloves used to prevent dangerous chemicals penetrating the skin

B For safety harnesses used to prevent or arrest falls from height

C For hearing protection to prevent exposure to harmful levels of noise

D For head protection from falling objects or hitting the head

E For high-visibility clothing worn in the proximity of moving traffic

6.54 As required by the Personal Protective Equipment at Work (Amendment) Regulations 2022, your site management team needs to carry out an assessment to determine whether the personal protective equipment (PPE) you intend to provide will be suitable. Which one of the following common items of PPE should also be risk assessed for suitability under the requirements of other health and safety regulations?

A A high-visibility jacket worn to minimise the risks of contact with vehicles or when moving plant

B A construction safety helmet used to protect against risks from falling objects or banging the head

C A pair of safety gloves used to prevent the risks of dangerous chemicals penetrating the skin

D A pair of safety boots worn to prevent injuries from dropped objects and penetration of the sole

6.55 In which one of the following circumstances is it appropriate to provide personal protective equipment to employees?

A Where there are significant risks identified to an individual's health and safety

B When adequately controlling the risks involved without increasing the overall level of exposure

C Where the conditions involved increase the likelihood of exposure to uncontrolled hazards

D As the last resort after all other risk methods and controls have been considered

6.56 Providing personal protective equipment (PPE) should be one of the first considerations after discovering a significant risk to the health of your employees. Is this correct?

A. Yes - PPE should be one of the first considerations for tackling the immediate risk, while other control measures are being developed

B. No - PPE should only be considered where a risk has been identified, and it cannot be adequately controlled by other means

C. Yes - PPE should be a primary consideration for controlling significant risks, in conjunction with technical and organisational means

D. No - PPE should be only considered as a short-term control measure, as opposed to eliminating the significant risk at source

6.57 Under the requirements of the Personal Protective Equipment at Work (Amendment) Regulations 2022, in order for personal protective equipment to be deemed as suitable, it must meet which of the following criteria?

1	It takes account of ergonomic requirements	A	1 and 2 only
2	It fits after making the necessary adjustments	B	1 and 3 only
3	It takes account of individual characteristics	C	2 and 3 only
4	It is freely available in a range of different sizes	D	2 and 4 only

6.58 Which **two** of the following criteria must be met in order for personal protective equipment (PPE) to be deemed suitable for use?

A. It is chosen based on the environment and surrounding conditions

B. It achieves the maximum assigned levels of protection in practice

C. It is provided on an individual basis and cannot be shared with others

D. It gives maximum protection while ensuring minimum discomfort

E. It effectively eliminates the specific risks for which it has been provided

6.59 When conducting face-fit testing for respiratory protective equipment (RPE) for a number of operatives on site, which of the following are significant factors in relation to RPE performance?

1	That the operatives are clean-shaven, to ensure respirator seals are not compromised	A	1 and 4 only
2	That the operatives' face shapes are considered in the selection of a respirator	B	1 and 3 only
3	That the operatives are given clear instructions on respirator storage and maintenance	C	2 and 3 only
4	That the operatives' medical histories are considered, due to the nature of the test	D	1 and 2 only

6.60 All workers are required to wear mandatory personal protective equipment (PPE) on site. How can an employer demonstrate 'good practice' when trying to persuade workers to wear PPE at all times?

A By supplying a safety data sheet on the benefits of PPE equipment

B By telling workers in the site safety induction that this is a site rule

C By involving the workers in the process of selecting the PPE purchased

D By asking the workers to complete a survey on their preferred choice of PPE

6.61 Work is being planned on the inside of an open-topped inspection chamber. Due to the nature of the work and the environment, it has the potential to become an oxygen-deficient atmosphere. Certain types of respirators are suitable to be used for this type of work. Is this correct?

A Yes - ventilated visors and helmet respirators can be used in oxygen-deficient atmospheres

B No - breathing apparatus must be used where there is the possibility of reduced oxygen levels

C Yes - powered respirators are suitable to be used where there is the possibility of reduced oxygen levels

D No - breathing apparatus must only be used where there is the possibility of oxygen-enriched atmospheres

6.62 You are due to start work on a project and your employer has asked you to purchase your own personal protective equipment (PPE). Which **two** of the following are significant regarding this request?

A You must not pay for PPE as this is an employer's duty

B You must pay for the replacement of damaged PPE

C You must pay for the replacement of the same item of PPE

D You must be provided with suitable PPE as stated in the risk assessment

E You must always pay a contribution towards PPE

6.63 When selecting personal protective equipment (PPE) which one of the following principles must be considered first?

A It must be supplied with information of how to use it safely

B It must be appropriate for the risks and exposure conditions

C It must not affect the performance of the worker

D It must be adjustable so that it provides a correct fit

6.64 Which one of these statements about personal protective equipment (PPE) is false?

A Workers must pay for any damage or loss

B Workers must store it correctly when it is not in use

C Workers must report any damage or loss immediately

D Workers must use it as instructed and look after it

6.65 Which one of the following must provide personal protective equipment to their workers?

A The employer

B The designer

C The client

D The project manager

6.66 In line with the Personal Protective Equipment (PPE) at Work Regulations which one of the following must workers comply with?

A Maintain PPE in good condition and replace it when it becomes defective

B Inform your supervisor of any damage to PPE and replace it at the end of the week

C Wear PPE every day as access to the site will be denied in line with site rules

D Supply new PPE every day to offer maximum protection

07 Dust and fumes (Respiratory hazards)

7.01 Which of the following is the most effective way of preventing or eliminating the presence of respiratory hazards on a construction project?

A Providing suitable protective equipment to all workers on site to control residual risk

B Ensuring there is a dedicated cutting area provided to avoid dust getting into the air

C Using water suppression on all cutting equipment that generates or creates dust

D Planning for the manufacturing or cutting of materials at a location away from site

7.02 Which one of the following control measures would eliminate air-borne dust from a circular saw?

A Attaching a suitable extraction unit

B Opening the windows in the area

C Fitting a wider cover to the blade

D Fixing a water suppression system

7.03 Which one the following can be implemented to prevent respiratory hazards facing a gang of bricklayers?

A The introduction of block splitters

B The use of respiratory protective equipment

C The implementation of task rotation

D The increase in health surveillance

7.04 Identify **two** management principles that will control dust hazards in the workplace.

A Any expected exposure must be controlled to an acceptable level

B All employees are informed of all control measures in place

C Risk assessments are completed at least two days in advance

D All work is done outside in the open to ensure adequate ventilation

E All new starters are informed about the health effects of dust exposure

7.05 Identify **two** potential risks that atomised aerosols can cause in the workplace if not properly controlled.

A They may carry bacteria which can cause harm to the person exposed

B They can cause serious harm like legionnaires disease

C They are toxic by inhalation causing unconsciousness and death

D They have chronic effects such as causing liver damage or sensitisation

E They give off vapours at room or low temperatures which are highly volatile

7.06 It is important to control wood dust when cutting or sanding wood because it can cause...

A occupational asthma and nasal cancer

B emphysema and lung elasticity loss

C silicosis and scarring of the lung

D pleural thickening and pulmonary disease

7.07 Construction dusts can cause irreversible health conditions through regular and uncontrolled exposure. Is this correct?

A Yes. These illnesses normally take many years to become visible and can be paralysing or fatal

B No. Sites manage and control operations that create dust to minimise exposure to workers

C Yes. Exposure can cause a range of conditions, including effects such as chronic coughing

D No. Personal protective equipment is issued to all workers to prevent uncontrolled exposure

7.08 You are chasing out walls inside a small room. Identify two control measures to prevent workers from exposure to airborne dust.

1	Use the correct on-tool dust extraction/water suppression and wear a high efficiency face mask like FFP3	A	1 and 2 only
2	Open the windows and door to allow the dust to escape into the atmosphere this removes the need for a mask	B	1 and 3 only
3	Use the on-tool dust extraction/water suppression and keep everyone else from entering the room to prevent exposure	C	1 and 4 only
4	Open the door and windows and wear a face mask when dust starts to build up to prevent or reduce exposure to dust	D	2 and 4 only

7.09 Identify **two** control measures that will help to prevent exposure to diesel fumes in an area where free movement of air is restricted and there is no ventilation.

A Workers must wear personal monitors to warn of carbon monoxide build up

B A banksman must be present to warn others of the build-up of carbon monoxide

C A supervisor must be present, wearing personal monitors to control carbon monoxide

D Air monitoring in the building must be in place to warn of carbon monoxide build up

E A risk assessment must clearly specify the acceptable levels of carbon monoxide

7.10 A worker is using a petrol disc cutter to cut concrete paving slabs. The disc cutter has a suitable dust suppression water bottle, and the operator is not clean-shaven, but is wearing a face mask. Is this adequate protection?

A Yes - concrete dust contains very heavy particles, which fall to the ground and cannot enter the airway through a small gap in the mask

B No - a worker could still be exposed, as facial hair only protects against larger dust particles that can be seen by the human eye

C Yes - using the dust suppression water bottle will eliminate the risks to the operator from harmful levels of exposure to dust

D No - facial hair can create an inadequate fit, which will significantly reduce the protection provided to the wearer

7.11 When conducting a welding fabrication task that involves heating up metal, an occupational health risk is created as the operative could...

A. lose their vision permanently from the bright arc created

B. inhale fine metal particles causing severe burns to their respiratory tract

C. burn their skin with hot metal fragments from the flashback created

D. inhale microscopic metal particles created by the fumes cooling down

7.12 If a young construction worker was exposed to low levels of silica dust over a period of time and has since left the construction industry, the effects of exposure will continue to develop long after the exposure has stopped, and it is irreversible. Is this correct?

A. No. Sufferers are likely to have severe shortness of breath and the symptoms will improve over time

B. Yes. Sufferers are not likely to be diagnosed until years after the exposure to silica has occurred

C. No. Sufferers usually have breathlessness and a persistent cough with phlegm which improves over time

D. Yes. Sufferers usually become house- or bed-bound and often die prematurely due to heart failure

7.13 One reason why an existing gas installation must be inspected by a gas safe engineer following any refurbishment work is to prevent what type of poisoning?

A. Carbon disulphide

B. Carbon dioxide

C. Carbon hydroxide

D. Carbon monoxide

7.14 Which one of the following are the two main types of silicosis workers can suffer from after high levels of exposure to silica dusts in the construction industry.

A. Accelerated and acute silicosis

B. Atrophied and chronic silicosis

C. Acute and chronic silicosis

D. Acute and pleural silicosis

7.15 What must your employer provide concerning silica dusts?

A. Information on when to use water suppression as a control measure

B. Product safety sheets on how to manage and prevent exposure

C. Toolbox talks on how to use the tool extraction as a safety measure

D. Employee training on the hazards resulting from any exposure

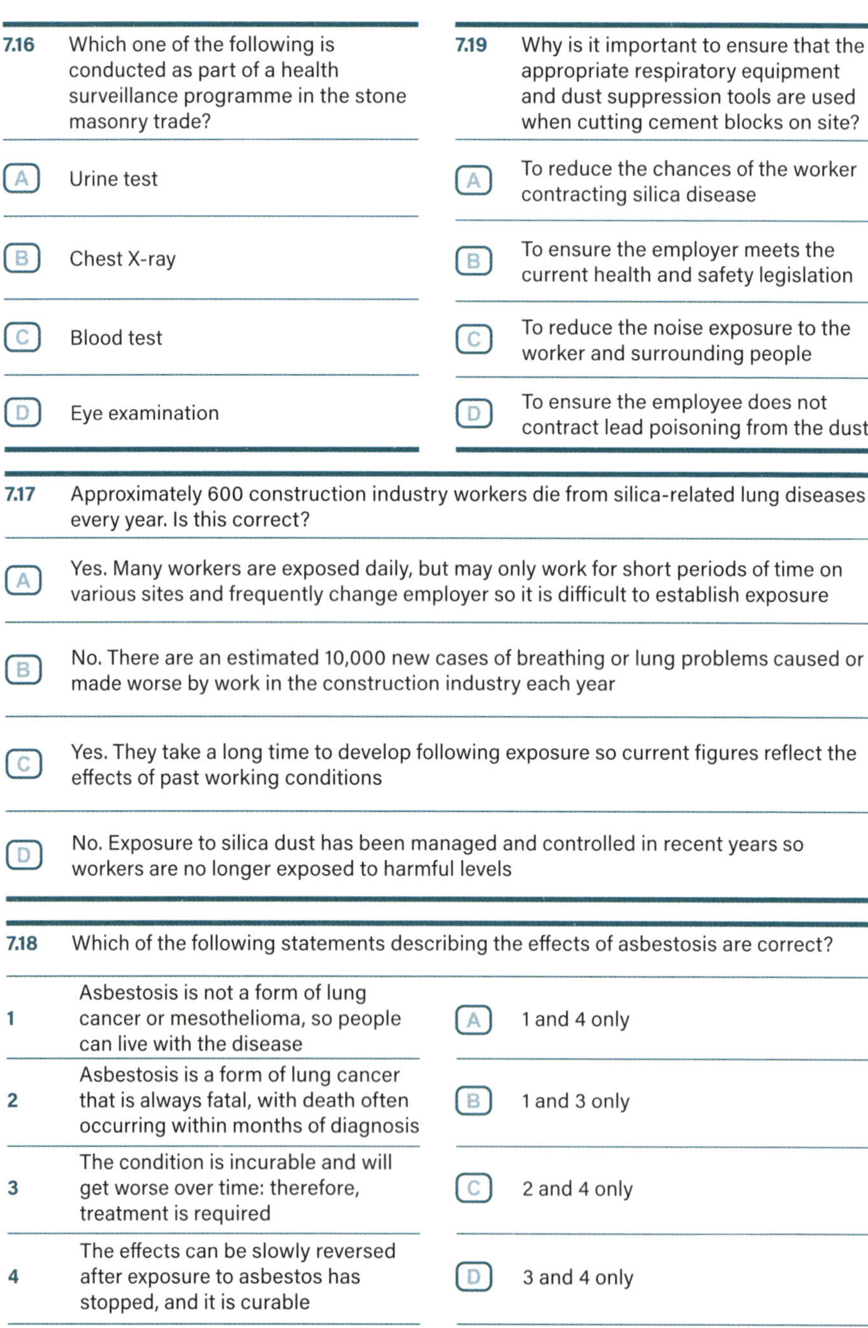

7.16 Which one of the following is conducted as part of a health surveillance programme in the stone masonry trade?

A Urine test

B Chest X-ray

C Blood test

D Eye examination

7.19 Why is it important to ensure that the appropriate respiratory equipment and dust suppression tools are used when cutting cement blocks on site?

A To reduce the chances of the worker contracting silica disease

B To ensure the employer meets the current health and safety legislation

C To reduce the noise exposure to the worker and surrounding people

D To ensure the employee does not contract lead poisoning from the dust

7.17 Approximately 600 construction industry workers die from silica-related lung diseases every year. Is this correct?

A Yes. Many workers are exposed daily, but may only work for short periods of time on various sites and frequently change employer so it is difficult to establish exposure

B No. There are an estimated 10,000 new cases of breathing or lung problems caused or made worse by work in the construction industry each year

C Yes. They take a long time to develop following exposure so current figures reflect the effects of past working conditions

D No. Exposure to silica dust has been managed and controlled in recent years so workers are no longer exposed to harmful levels

7.18 Which of the following statements describing the effects of asbestosis are correct?

1 Asbestosis is not a form of lung cancer or mesothelioma, so people can live with the disease

2 Asbestosis is a form of lung cancer that is always fatal, with death often occurring within months of diagnosis

3 The condition is incurable and will get worse over time: therefore, treatment is required

4 The effects can be slowly reversed after exposure to asbestos has stopped, and it is curable

A 1 and 4 only

B 1 and 3 only

C 2 and 4 only

D 3 and 4 only

7.20 What health condition might you develop if you work in a dusty environment over long periods of time?

A Occupational asthma

B Occupational dermatitis

C Skin cancer

D Weil's disease (leptospirosis)

7.21 Which of the following statements is the most accurate description of the harm that may be caused by exposure to solvent vapours?

A They deposit bacteria into the body at locations where the most harm can be caused

B They can often be toxic, causing chronic effects or dizziness, unconsciousness and death

C They are hazardous in enclosed spaces, as they quickly build up to dangerous concentrations

D They can cause a range of conditions, from coughing and sneezing to chronic disease

7.22 After knowingly or unknowingly breathing in silica particles over a long period of time, which of the following conditions are construction workers most likely to suffer from due to exposure to low levels of respirable crystalline silica (RCS)?

A Chronic effects such as liver damage

B Hardening or scarring of the lung tissue

C Severe or chronic occupational asthma

D Diffuse pleural thickening of the lung

7.23 Which **two** of the following work processes on site can potentially expose workers to respiratory sensitisers that can cause occupational asthma?

A Spray painting using two-pack paints

B Cutting slabs, kerbs or concrete

C Second fix joinery and carpentry work

D Oxy-acetylene or oxy-propane welding

E Plant powered by internal combustion engines

7.24 You are about to start work in a confined space, and you will be using an engine-powered generator (110 volts) to power your tools. Which **two** of the following may be acutely fatal as well as causing long-term consequences such as lung cancer?

A Carbon dioxide produced by the exhaust polluting breathable air

B Unburnt hydrocarbons produced by incomplete combustion of the exhaust

C Hydrogen monoxide produced by engine batteries when the generator is running

D Carbon monoxide produced by restriction on the movement of air

E Dihydrogen monoxide producing vapours as the exhaust is heating up

08 Noise and vibration

8.01 What might be the health outcome for an individual exposed to noise in the workplace?

A Permanent and disabling hearing damage

B Temporary inner-ear imbalance and vertigo

C Long-term chronic earache and infections

D Continually prone to temporary threshold shift

8.02 Which one of the following could be an early indication of hearing damage?

A Difficulty having conversations where there is background noise

B Inability to hear vehicle and plant audible warnings

C Needing to increase the setting level of hearing protection

D Having to take longer breaks due to the onset of headaches

8.03 Which **two** of the following are true in relation to hearing damage?

A It is irreversible and will not improve

B It can be reversed if caught early enough

C It is only caused by noisy working environments

D It is compounded by the ageing process

E It can be treated with ongoing medication

8.04 Which one of the following is a consequence for an individual's hearing when exposed to low-level noise over many years?

A Social conversations with a group of people can become difficult

B Psychological problems can develop, sometimes suddenly

C Damage will happen slowly, go unnoticed and result in significant harm

D Some sounds appear identical and this can lead to unclear speech patterns

8.05 A large construction project has identified that hearing protection must be worn in certain build sections, and will become a mandatory policy requirement. What are the drawbacks of this policy?

A It will compromise the productivity of workers in these areas

B It will take too much management time to ensure compliance

C It will increase project costs outside of the allocated budget

D It will decrease workers' awareness of all other hazards

8.06 A joinery workshop has a number of old mortise and tenon drilling machines. What could be the main health impact for workers operating these machines if control measures are not in place?

(A) Long-term use could have a cumulative effect on the workers' hearing due to the noise of the working parts

(B) Long-term use could potentially affect the workers' eyesight due to the small intricate parts of the machine

(C) Long-term use could significantly affect overall productivity of the workers, compromising their stress levels

(D) Long-term use could negatively affect the workers' postural positioning, leading to lower back and leg strain

8.07 When visiting a site your team needs to complete an inspection that is adjacent to an activity generating a high level of noise, which cannot be stopped. Which one of the following actions would you expect the site manager to apply?

(A) Arrange for a noise assessment to be conducted before the inspection can start

(B) Make hearing protection available to those who ask for it

(C) Issue everyone affected with hearing protection and ensure it is worn

(D) Give a toolbox talk on how to work safely in a noisy working environment

8.08 If you are suffering from whistling and buzzing in your ears, it is nothing to be worried about as it is a consequence of working in a noisy construction environment, and these symptoms will subside. Is this true?

(A) Yes - this is only a temporary and short-term painful condition

(B) No - this is only treatable when exposure to noise stops permanently

(C) Yes - this is a common industry problem, and no permanent damage will be caused

(D) No - this is an early warning of hearing damage, and must not be ignored

8.09 Nuisance noise can cause individuals to suffer physiological effects, but it must not be compared to the effects of excessive or long-term exposure to noise. Is this true?

A Yes - nuisance noise is below the level that could damage a worker's hearing

B No - nuisance noise is a common cause of noise-induced hearing loss amongst workers

C Yes - nuisance noise causes severe hearing complications amongst workers

D No - nuisance noise can easily be controlled to remove the impact on workers

8.10 A recent health survey has identified that a number of construction operatives have tinnitus, and there are concerns about how the working environment might affect them. Which one of the following statements is correct?

A The hearing protection they were issued with will have to be changed to units that have been specifically manufactured to soothe the buzzing symptoms

B The job activities they usually undertake will have to be rotated, to ensure that they are only exposed to low levels of noise: otherwise, the buzzing symptoms will worsen

C They will need to take longer to complete their daily activities, as the buzzing symptoms will be heightened, which will affect their concentration levels

D The background noises of the site environment will help to mask the continual buzzing symptoms, so day-to-day activities will be manageable

8.11 To manage noise levels during construction activities, which one of the following hierarchical approaches should be adopted?

A Source, path, receiver

B Risk, source, control

C Source, selection, maintain

D Identify, manage, monitor

8.12 When it is not reasonably practicable to remove noise hazards, and a level of noise remains, which one of the following must be provided?

A Suitable training and awareness

B Suitable risk monitoring

C Suitable hearing protection

D Suitable acoustic barriers

8.13 When choosing suitable personal protective hearing equipment, which one of the following protection factors is important?

A It must ensure that all noise is continually eliminated

B It must eliminate risks from noise to an acceptable level

C It must magnify certain sound frequencies over others

D It must ensure comfort and durability when working

8.14 Which one of the following must an employer do if workplace noise levels are at or above the exposure limit value of 87 dB(A)?

A Conduct daily noise assessments to gather more data

B Issue hearing protection with higher attenuation levels

C Deliver training sessions on how to manage and control

D Identify and adjust organisational and technical measures

8.16 A health surveillance programme has identified that a number of site-based employees have suffered hearing loss. Which **two** of the following actions must the employer now take?

A Repeat the audiometry test again in six months

B Advise the employees to speak to their doctors

C Refer them to a doctor or specialist for examination

D Review all the noise-related risk assessments

E Maintain the surveillance records and disclose if required

8.15 You are standing in a busy section of the site with the quantity surveyor. You are stood approximately one metre away from each other, but are having to raise your voices to hold a conversation. Why is this happening?

A The quantity surveyor has hearing aids in both ears, and this is being accounted for as the noise level is only 70 dB(A)

B You are standing too far apart, and need to be much closer to engage in clear conversation, as the noise level is around 85 dB(A)

C Hearing protection is being worn, but the ear protectors have been damaged, and the noise level is above 87 dB(A)

D Hearing protection is not being worn in line with the risk assessment for this area, and the noise level is at 90 dB(A)

8.17 On a construction site where noise levels are not able to be practicably reduced below the upper exposure action value (UEAV), and where employees are likely to be exposed, what must be put in place to mitigate the impact of noise?

A. A programme of practical workshops demonstrating how to fit and wear hearing protection correctly

B. Designated hearing protection zones and suitable signage stating that hearing protection must be worn

C. Provision of suitable hearing protection, if a worker requests it according to their own personal preference

D. Appropriate work rotation aligned with sufficient rest breaks, so that hearing protection is not needed

8.18 In line with the Control of Noise at Work Regulations 2005, a reliable and representative estimate of workers' daily personal noise exposure must be made. Which one of the following states how this is estimated?

A. Through a combination of how the noisy work is done and how long each worker takes to do the same task

B. Through a combination of how loud and for how long a worker is exposed to noise in a working day

C. Through an analysis of daily noise measurements conducted while workers do specific tasks

D. Through an analysis of calculated peak sound pressure levels that a worker is exposed to in a working week

8.19 Which of the following measures will provide continual medium and long-term benefits as part of a construction project's noise control programme?

1	Investing in a 'buy quiet' policy for machinery and equipment	A	1 and 2 only
2	Improving the working techniques to reduce noise levels	B	2 and 3 only
3	Erecting enclosures using absorbent materials to reduce the noise	C	3 and 4 only
4	Maintaining noise-omitting machinery and equipment on a regular basis	D	1 and 4 only

8.20 When referring to manufacturer's noise data about equipment, it is the employers responsibility to ensure further information on how it will be used in the workplace is provided. Is this correct?

A Yes, because the manufacturers data provides only guidance on personal noise exposure, and there are many other factors that affect noise levels

B No, the data provided by the manufacturer is extremely accurate as they conduct a range of noise tests to ascertain personal noise exposure levels

C Yes, as the data must be representative of how the equipment will be used so appropriate training and awareness programmes can be developed

D No, as manufacturers must comply with the Provision and Use of Work Equipment Regulations 1998 and always provide accurate noise exposure data

8.21 Which one of the following is a symptom of hand-arm vibration syndrome (HAVS)?

A Uncontrolled movement in the wrists and fingers

B Contagious blistering on the hands and arms

C Increased level of dry skin between the fingers

D Reduced strength in the hands and fingers

8.22 Why is overexposure to vibratory tools and equipment a serious issue?

A Its impact and severity on long-term health is still unknown

B There are no early warning symptoms so damage cannot be mitigated

C The associated health conditions are difficult to assess and diagnose

D It can cause permanent and disabling health conditions

8.23 Which one of the following will trigger the vascular symptoms of hand-arm vibration syndrome?

A Frequency of tool use

B Posture of the tool user

C Working in cold temperatures

D Limited manual dexterity

8.24 When using a disc cutter, which one of the following might increase the vibration levels?

A The user maintaining a tight grip on the tool

B The tool being operated on its highest setting

C The tool's disc always being sharp and clean

D The user being relaxed whilst operating the tool

8.25 Following a health surveillance programme conducted on site, the analysis of results has indicated that a number of workers are suffering from musculoskeletal symptoms linked to hand-arm vibration syndrome (HAVS). Which one of the following conditions would have led to this conclusion?

A Loss of sensation in the fingers and hand

B Degeneration of the upper-arm rotator cuff

C Blanching of one or all of the fingers

D Limited, or loss of, manual dexterity in the hands

8.26 Operatives who have been diagnosed with stage 2 hand-arm vibration syndrome (HAVS) have been given advice on how to minimise their vascular symptoms. Which **two** of the following will help the operatives to control these symptoms?

A To stop cigarette smoking, or cut down on intake

B To take more rest breaks during the working day

C To keep warm and dry in cold and damp weather

D To wear anti-vibratory gloves when appropriate

E To hold the correct posture when operating a tool

8.27 You observe an operative needing to apply considerable pressure while operating a road-breaking drill to break up part of a road. Which one of the following effects of this activity, and ensuing health risk, would occur?

A The noise levels of the tool, affecting their hearing sensitivity

B The sparks created by the tool, resulting in third-degree burns to their skin

C The vibrations through the tool, creating hand-arm vibration syndrome

D The heat created by the tool, creating a short circuit and giving them a minor shock

8.28 To help control the effects of hand-arm vibration syndrome (HAVS), the exposure action value (EAV) and the exposure limit value (ELV) need to be adhered to. Which one of the following explains how they are measured?

A By using the tool manufacturer's vibration exposure data to establish an accurate trigger time limit for workers

B By observing workers using vibratory tools and equipment, and using the data gathered to manage the exposure limits

C By analysing a combination of the vibration at the grip point on the tool and the time a worker spends gripping the tool

D By calculating the tool's vibratory output at varying speeds, compared with the hours a worker has spent operating it

8.29 Which of the following are neurological hand-arm vibration syndrome (HAVS) symptoms?

1	Numbness and tingling in the fingers	A	1 and 2 only
2	Reduced sense of touch and temperature	B	2 and 3 only
3	Muscle fatigue and loss of grip	C	3 and 4 only
4	Finger blanching and red throbbing	D	1 and 3 only

8.30 It is recognised that symptoms of hand-arm vibration syndrome (HAVS) may be intermittent, but with continual exposure to vibration, they could become prolonged or even permanent. Is this true?

A Yes - and if the symptoms become permanent, it is very unlikely that they will cause any physical pain or psychological impact

B No - the onset of symptoms is immediate, and there is no prior warning for the individual to try and minimise the impact

C Yes - continual exposure can become a permanent condition, and will cause pain, distress and sleep disturbance

D No - although the symptoms come and go, it would only be temporary as it is treatable with medical intervention

8.31 Which one of the following should be implemented when it is not possible to remove vibration from a work process?

A Adoption of a purchase or hire smooth policy

B Limitation on individuals daily trigger time

C Encouraging individuals to disclose symptoms

D Issuing of anti-vibration gloves

8.32 A health surveillance programme in place for workers who are exposed to vibration will need to include individuals who have been diagnosed with hand-arm vibration syndrome, even when exposed to levels below the exposure action value. Is this correct?

A Yes - it is an employer's duty under the Control of Vibration at Work Regulations 2005

B No - as the upper exposure action value is not being exceeded, the risk is controlled

C Yes - as it is more cost-effective to include this group of workers in vibration health surveillance

D No - it is medically assessed and monitored outside of the Control of Vibration at Work Regulations 2005

8.33 Which one of the following is the most effective way of controlling vibration exposure on a construction site?

A Using equipment with low vibration output

B Limiting the operatives trigger time

C Maintaining equipment so it remains efficient

D Changing the process and avoidance

8.34 When a risk assessment has identified that workers are at risk from vibration exposure, which one of the following must their employers implement?

A A series of toolbox talks to explain vibration-control measures and symptom awareness

B A control programme to reduce exposure to below the exposure limit value

C A safe system of work based on tool rotation and trigger-time durations

D A data analysis programme to control all exposures above the action value

8.35 When an operative has reached their agreed vibration exposure points level, which one of the following must they do?

A Speak to their supervisor, so that they can be allocated work that does not involve vibrating hand tools

B Inspect their vibrating hand tool for damage, and then ensure that it is correctly stored before next use

C Change their hand tool for a less-vibratory tool that doesn't require exposure points to be allocated

D Log their vibration exposure points on the central recording system, and leave site for the day

8.36 Following health surveillance on site, it has been established that a number of operatives are being negatively affected by vibration. Which **two** of the following must the employer now do?

A Look at the practicalities of reassigning the operatives to other work that does not involve vibration

B Review current control measures in line with any advice given by the occupational health professional

C Speak to the operatives and explain their results, and that they will be re-tested in three months

D Remove the operatives from vibratory work tasks for one month, and then gradually re-introduce them

E Speak to all other operatives so they can be made aware of the consequences of vibration exposure

8.37 A group of operatives are using hand-held breakers to break up a pavement ready for landscaping. How can this activity be controlled to ensure that the health of the workers is not compromised?

A Check that the equipment has been regularly serviced before use, and that any vibration-resilient components have been replaced

B Check that the operatives keep a diary recording their daily tool output, in order to protect the company from future claims

C Check that, as part of the risk assessment process, individual vibration exposure levels are matching the manufacturer's data

D Check that the operatives are issued with anti-vibration gloves, and that these are worn when operating the equipment

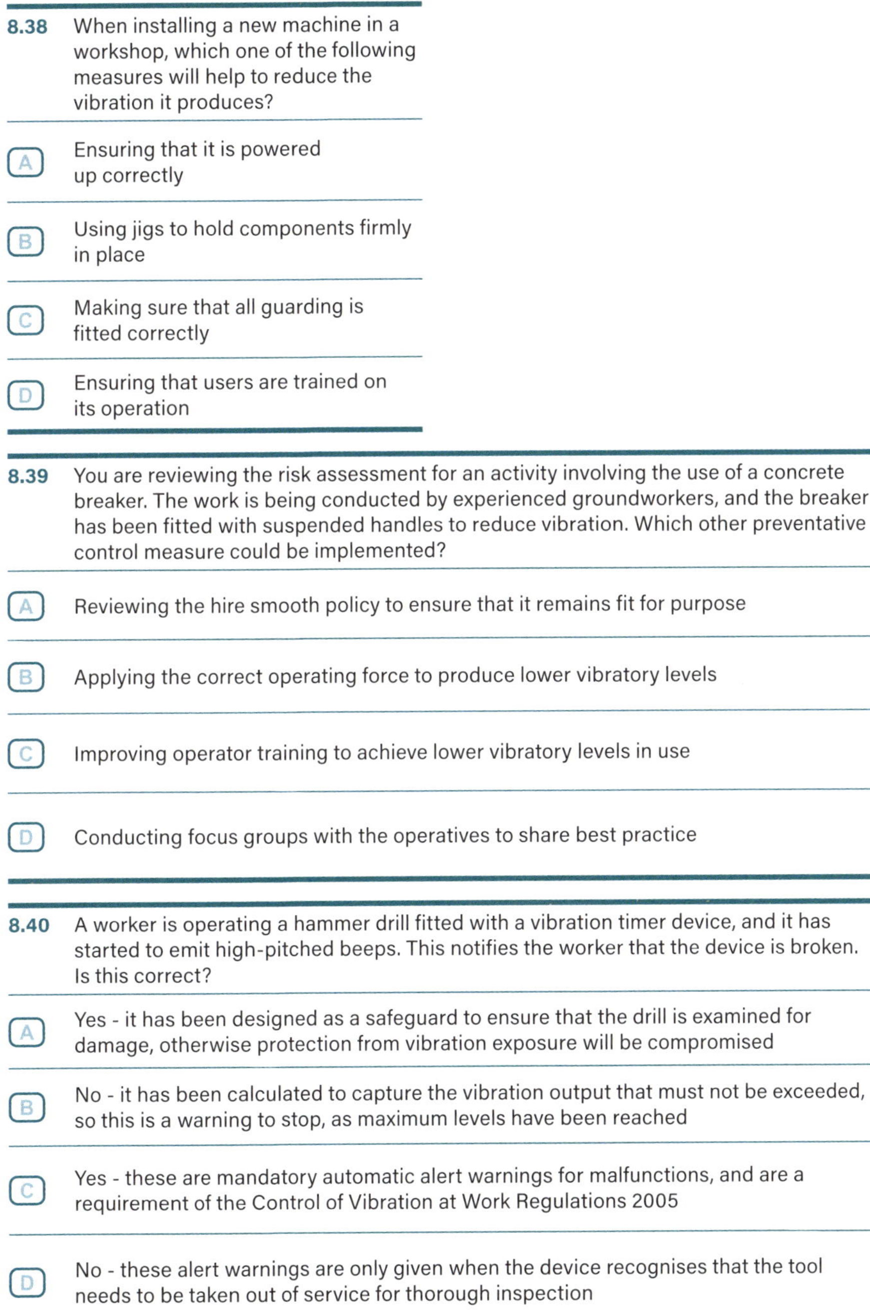

8.38 When installing a new machine in a workshop, which one of the following measures will help to reduce the vibration it produces?

A) Ensuring that it is powered up correctly

B) Using jigs to hold components firmly in place

C) Making sure that all guarding is fitted correctly

D) Ensuring that users are trained on its operation

8.39 You are reviewing the risk assessment for an activity involving the use of a concrete breaker. The work is being conducted by experienced groundworkers, and the breaker has been fitted with suspended handles to reduce vibration. Which other preventative control measure could be implemented?

A) Reviewing the hire smooth policy to ensure that it remains fit for purpose

B) Applying the correct operating force to produce lower vibratory levels

C) Improving operator training to achieve lower vibratory levels in use

D) Conducting focus groups with the operatives to share best practice

8.40 A worker is operating a hammer drill fitted with a vibration timer device, and it has started to emit high-pitched beeps. This notifies the worker that the device is broken. Is this correct?

A) Yes - it has been designed as a safeguard to ensure that the drill is examined for damage, otherwise protection from vibration exposure will be compromised

B) No - it has been calculated to capture the vibration output that must not be exceeded, so this is a warning to stop, as maximum levels have been reached

C) Yes - these are mandatory automatic alert warnings for malfunctions, and are a requirement of the Control of Vibration at Work Regulations 2005

D) No - these alert warnings are only given when the device recognises that the tool needs to be taken out of service for thorough inspection

OCCUPATIONAL HEALTH, WELLBEING AND WELFARE

09 Hazardous substances

9.01 Plumbers are removing lagging around pipes to repair a leak in a Victorian building. What is the significant risk to workers' health in carrying out this work?

A They are disturbing asbestos fibres that can be inhaled

B They are releasing flying fragments that can injure a person

C They are disturbing bacterial-filled water that can cause infection

D They are disturbing the dust gathered that can irritate eyes

9.02 When asbestos fibres are breathed in at a young age and no further exposure occurs, this allows the lungs to recover and reduces the development of an asbestos related disease. Is this true?

A Yes, this will allow enough time for the lung and chest lining tissues to fully recover

B No, timescales of exposure to fibres have no relevance in the development of future health disease

C Yes, it depends on the type of asbestos fibres inhaled as to whether a future health issue will develop

D No, inhaling fibres at an early age greatly increases the chances of developing an incurable disease

9.03 Exposure to asbestos can cause which of the following **three** related health diseases?

A Asbestosis

B Bronchitis

C Silicosis

D Mesothelioma

E Pleural thickening

9.04 What illness could a worker develop if they breathe in asbestos dust?

A Weil's disease

B Pneumoconiosis

C Mesothelioma

D Occupational asthma

9.05 You have been asked to carry out alterations on a building built before the year 2000. What would be your first consideration?

A The building could contain fibreglass insulation

B The building could contain asbestos-based materials

C The building's finishings could be painted with lead paint

D The internal walls could be made from lath and plaster

9.06 Why is it important to ensure all workers attend asbestos awareness training?

(A) To ensure workers meet the company's asbestos-related health and safety standards

(B) To enable workers to identify potential asbestos-containing materials and to know what to do

(C) To give workers enough knowledge to be able to remove the asbestos-containing materials

(D) To keep workers training records up-to-date and in line with legislative information

9.07 An electrician needs to run cables to create a new lighting circuit over asbestos sheeting in poor condition. What must be followed to ensure compliance for this type of work?

(A) This is notifiable non-licensed work and specific processes must be followed

(B) This type of asbestos material must be subject to sampling by a specialist

(C) This is licensed work and must be carried out by a qualified registered company

(D) This type of work must comply with current electrical standards and be certified

9.08 When an asbestos refurbishment survey needs to be conducted it should be based on the...

(A) project's schedule

(B) access limitations

(C) project's requirements

(D) surveyor's fees

OCCUPATIONAL HEALTH, WELLBEING AND WELFARE

9.09 Which of the following is true regarding the risks from asbestos?

1	The earlier the date of first exposure, the greater the chance of recovery	A	1 and 2 only
2	The most friable materials need higher levels of control	B	2 and 3 only
3	Mesothelioma can be curable if treated with chemotherapy	C	3 and 4 only
4	If not disturbed or damaged it may not need removing	D	2 and 4 only

9.10 Why are cement-, mortar- and concrete-based materials hazardous to health?

A They can cause dizziness and headaches

B They can cause skin burns and irritant dermatitis

C They can cause muscle aches and spasms

D They can cause skin sores and abrasions

9.11 You have observed that the site mechanic carrying out repairs to the hydraulics of the 360° excavator is not wearing the correct protective gloves. What significant dangers are they being exposed to?

A They are increasing their chances of having an anaphylactic shock

B They are increasing their chances of getting Raynaud's disease

C They are increasing their chances of getting skin cancer

D They are increasing their chances of nerve damage to the fingers

9.12 Which one of the following is the least likely to cause skin problems?

A Asbestos

B Bitumen

C Cement

D Concrete

9.13 Site operatives have finished laying slabs, and need to wash out the buckets that have stored the wet concrete. Which one of the following actions could they take which would not reduce their exposure to burns from this process?

A Using the correct cleaning agent along with hot water

B Reducing the total amount of time taken to carry out the task

C Using suitable washing and changing facilities afterwards

D Wearing appropriate personal protective equipment

9.14 Why do cement and concrete burns pose such a serious risk to workers' health?

A They can damage skin nerve endings very quickly

B They can increase body temperature rapidly

C They can cause a loss of consciousness

D They can spread to other parts of the body

9.15 Workers are washing out cement buckets on site. What is the main risk to their health?

A Respiratory difficulties

B Skin cancer

C Flu-like symptoms

D Burns to skin

9.16 Which of the following **two** main health risks can be caused by working with wet cement and concrete?

A Headaches

B Dizziness

C Skin burns

D Dermatitis

E Abrasions

9.17 Workers are screeding a wet concrete floor by hand. Which two of the following will help to reduce the potential health affects from this activity?

1	Portable air extraction units	A	1 and 2 only
2	Appropriate wellington boots	B	2 and 3 only
3	Suitable washing facilities	C	3 and 4 only
4	Rubber handled tools	D	1 and 3 only

9.18 You are delivering a health and safety briefing to operatives involved in a scheduled concrete pour. Does this work carry a risk to the operatives' health?

A Yes - wet concrete can cause severe burns that can kill nerve endings, resulting in long-term damage

B No - wet concrete can cause skin irritation and redness, but this is only temporary and it will clear up

C Yes - wet concrete can cause burns to the skin's surface, but these are easily healed with medication

D No - wet concrete is an alkali-containing substance, so it is safe if contact is made with the skin

9.19 Painters are using the correct process and suitable personal protective equipment to remove old lead paint work within a domestic property. What is the next main priority?

A Ensure workers have a suitable location for rest breaks outside of the working area

B Ensure property owners do not enter the working area until complete and the area cleaned

C Ensure there are suitable and sufficient welfare facilities for the workers present

D Ensure barriers and signage are erected to stop any unauthorised access to the working area

9.20 Which **three** construction activities are most likely to result in a worker being exposed to lead?

A Stripping old paint from windows and doors

B Disturbing sprayed coatings

C Repairing old gutterings

D Hot-cutting during demolition

E Cutting slabs, kerbs and concrete

9.21 It is regarded as good practice to remove old paint work by using a gas torch. Is this true?

A No, this type of material is highly resistant to flame and only sanding can remove it

B Yes, this is the safest and quickest way to remove this toxic material from timber surfaces

C No, this process should never be used on this material because it releases toxic fumes

D Yes, this allows all the material to be fully removed and reduces the need for sanding

9.22 Which one of the following tasks would not expose a worker to lead fumes?

A Burning off old paint with a hot-air gun

B Soldering heating pipes

C Removing old loft insulation

D Working inside tanks that contained petrol

9.23 Which one of the following could be a consequence to roofers who regularly repair lead flashings on properties?

A Long-term bronchitis through prolonged periods of working outdoors

B Long-term musculoskeletal disorders due to manual handling

C Long-term health issues relating to the exposure to asbestos

D Long-term health issues relating to the exposure to a toxic material

9.24 Painters are preparing paintwork in an old property by dry sanding. What is the significant risk to health from this process?

A Exposure to flying fragments

B Exposure to excessive noise

C Exposure to prolonged vibration

D Exposure to toxic dust

9.25 A painter is using a hot-air gun set at over 360°C to remove old paint from a timber frame window. What major health risk could they potentially be exposed to?

- A They could be at risk of inhaling asbestos fibres
- B They could be at risk of receiving third-degree burns
- C They could be at risk of inhaling lead fumes
- D They could be at risk of being exposed to heat radiation

9.28 Breathing in dust produced from using a masonry saw to cut concrete blocks slabs could result in which one of the following diseases?

- A Silicosis
- B Legionnaires'
- C Asbestosis
- D Leptospirosis

9.26 What are the two reasons why painters and decorators are advised to wet sand paintwork?

1. To produce a superior quality finish
2. To reduce the risk from cuts and abrasions
3. To stop possible toxic dust becoming air borne
4. To reduce the noise levels from the process

- A 1 and 2 only
- B 2 and 3 only
- C 3 and 4 only
- D 1 and 3 only

9.27 A risk assessment for demolition work has identified a low exposure and concentration level of lead-based materials. What additional steps must be put in place by the employer to reduce health risks?

- A Monitoring and recording the exposure levels of all workers
- B Displaying suitable warning signs highlighting the health risks
- C Cleaning all personal protective equipment after use
- D Stopping eating, drinking or smoking in contaminated areas

9.29 What is the next biggest health risk to construction workers after asbestos?

- A Leptospirosis
- B Silicosis
- C Legionnaires'
- D Bronchitis

9.30 Labourers are not following the safe system of work issued and have decided to dry sweep an area where stonemasons have been cutting sandstone blocks. Why is the process of dry sweeping creating a significant risk to their health?

A The sandstone dust can cause sensitising dermatitis

B The escape of silica dust can cause lung cancer

C The mixture of all ground dusts can cause respiratory asthma

D The bending action can cause musculoskeletal disorders

9.31 The **three** main effects that inhaling respirable crystalline silica (RCS) can have on a person's health is that they can develop...

A lung cancer

B heart valve disease

C chronic obstructive pulmonary disease

D deep vein thrombosis

E silicosis

9.32 Workers exposed to low levels of respirable crystalline silica (RCS) over a prolonged period will have no long-term health issues. Is this correct?

A Yes, this will have no effect on the person because the levels are minimal

B No, inhaling even a small amount of toxic material can affect a person's digestive system

C Yes, this long timeframe allows the body to recover and re-grow damaged tissue

D No, there is a risk that this will cause hardening and scarring of the lung tissue

9.33 Exposure to respirable crystalline silica (RCS) is harmful to a worker's health as it can...

A) increase the risk of kidney disease

B) cause deep burns and damage nerve endings

C) develop into acute asthma that is not reversible

D) cause reoccurring eye infections that require surgery

9.34 You are writing a risk assessment for the refurbishment of an old bathroom, and one of the main tasks to be controlled is the removal of tiles from the walls. Which one of the following, if not controlled sufficiently, would be the main health hazard of this work?

A) If the tiles are broken into sharp fragments, they could puncture the skin, causing hepatitis

B) If large dust clouds are created and then inhaled, there is a chance of contracting asbestosis

C) If bare skin makes contact with dirty tiles, then bacteria can be ingested, causing psittacosis

D) If fine dust particles are created and inhaled, there is a chance of contracting silicosis

9.35 A number of workers have been diagnosed with either tuberculosis, arthritis or kidney disease. Which of the following activities have they been undertaking over a number of years that have caused them to develop these health conditions?

1 Intricate stonemasonry work

2 Abrasive blasting of concrete

3 Removal of pipe lagging

4 Digging of excavations

A) 1 and 2 only

B) 3 and 4 only

C) 2 and 3 only

D) 1 and 4 only

9.36 You are writing the risk assessment for stonemasons who are working on a historical restoration project, which involves carving and removing rock, sand and clay. Which one of the following would you list as the main health hazard?

A The inhalation of fine dust when shaping materials can cause breathing difficulties and acute silicosis

B The moving and handling of materials into position can cause muscle fatigue and back injuries

C The constant hitting of materials with various hand tools can result in acute tendonitis

D The manipulation methods used to shape materials create fine fragments that can penetrate skin

9.37 Which one of the following is not a key stage of a control of substances hazardous to health (COSHH) risk assessment?

A Eliminate the use of the substance

B Identify who is at risk from the substance

C Erect suitable warning signs on storage areas

D Identify the hazardous substances

9.38 If it is not reasonably practicable to prevent exposure of workers to substances hazardous to health, which of the following must be considered first?

A The instruction, training and supervision required

B The use of sufficient control measures to control exposure

C The introduction of health surveillance arrangements

D The continual monitoring of workers exposure

9.39 Work procedures for decorators need to be changed from applying paint by hand, to using a spray application. What main document needs to be reviewed before these changes can be implemented?

A The provision and use of work equipment risk assessment

B The material safety data sheets

C The health surveillance guidance

D The control of substances hazardous to health risk assessment

9.40 The joiners on a project want to change their polyvinyl acetate (PVA) glue to a different supplier. What could happen if this is done before reviewing the current control of substances hazardous to health (COSHH) risk assessment?

A Workers may be exposed to new elements that can damage their health

B Workers may find that is not suitable for the purpose intended

C Workers could take longer to apply it as the consistency is different

D Workers could apply the PVA incorrectly as they have not had the suitable training

9.41 The mechanical engineer on site is planning to use a water-based hydrocarbon solvent for cleaning and degreasing plant components. Which **two** significant details must they be made aware of?

A The substance needs a licence to use

B The substance is corrosive

C The substance crystalises after application

D The substance is harmful

E The substance is highly flammable

9.42 A new degreasing agent is going to be used in a plant machinery workshop. Which one of the following needs to be considered before it is introduced?

A The workshop's control of substances hazardous to health safe systems must be reviewed before any further work is completed

B The competent person must review all the current control of substances hazardous to health risk assessments

C The health and safety manager must be informed to update the control of substances hazardous to health working procedures

D The current control of substances hazardous to health risk assessment must be reviewed by a competent person

9.43 Before a joiner can cut hard wood using a table saw, the task requires completion of a control of substances hazardous to health (COSHH) assessment. Is this true?

A No - this would only require an assessment of the safe use of the machinery

B Yes - the assessment would consider the dust levels created by the machinery

C No - this would only require an assessment of the electrical element of the machinery

D Yes - the assessment would consider the noise levels created by the machinery

9.44 You have been made aware that workers on site are using a highly-toxic cleaning agent unsafely, and you are not certain if the necessary paperwork is in place. Which two of these options are the first you must implement to reduce the risks to these workers?

1	Review the personal protective equipment (PPE) being worn by workers, and make them aware of the hazards	A	1 and 2 only
2	Erect appropriate barriers to stop any unauthorised access to the work area, and display warning signs	B	2 and 3 only
3	Stop the work immediately, ask the workers to leave the area and thoroughly ventilate the workspace	C	3 and 4 only
4	Instruct a competent person to carry out a control of substances hazardous to health (COSHH) risk assessment	D	1 and 4 only

9.45 You are writing the control of substances hazardous to health (COSHH) risk assessment for joiners using adhesives to fix materials on site. Which one of the following is not a significant factor when considering personal protective equipment as a risk control?

A It provides ultimate protection, and reduces the need for health surveillance checks

B It makes good occupational health and business sense to eliminate its purchase and use

C It only protects the wearer, and requires management efforts to ensure that it is used correctly

D It is expensive and unpleasant for the workers to wear, and bulk purchasing is required

9.46 Inhaling old mould spores during demolition work can expose workers to which one of the following diseases?

A Aspergillosis

B Leptospirosis

C Legionnaires' disease

D Silicosis

9.49 From the list below, select **three** ways that harmful substances can enter the human body.

A Absorption

B Ionisation

C Inhalation

D Metabolism

E Injection

9.47 A painter is using a highly-pressurised airless spray gun to apply paint to a large volume of walls and ceilings. Which one of the following is a major health risk of this work?

A The potential for a respiratory condition to develop from the release of hazardous fumes

B The potential for loss of hearing due to the noise levels omitted

C The potential for nerve damage to develop from the high level of vibration exposure

D The potential for loss of a finger, hand or arm due to fluid injection

9.48 Absorption of a hazardous substance happens when it...

A passes through the stomach after digestion, then into the blood stream

B passes through unbroken skin or cuts, then into the blood stream

C passes through the skin by piercing, then entering the blood stream

D passes through the lungs by breathing in, then filters into the blood stream

9.50 You are reviewing the current risk assessment for groundworkers repairing a section of a sewerage system. Why is it important to consider the personal protective equipment that they are wearing, and to ensure that they know the importance of maintaining good hygiene routines for this activity?

1	To reduce the risk of cross-contaminating canteen and office facilities	A	1 and 2 only
2	To reduce the risk of viruses spreading into common working areas	B	2 and 3 only
3	To reduce the risk of absorption of organisms through cuts and scratches	C	3 and 4 only
4	To reduce the risk of a needle-stick injury from a discarded hypodermic needle	D	1 and 4 only

9.51 A group of groundworkers are digging foundations for a new housing development on a brownfield site. Why is it significant to reinforce the importance to the workers of covering up any cuts or skin abrasions before starting work?

A They can be at risk of contracting swine flu, if wounds were to come into contact with stagnant water

B They can be at risk of contracting tetanus bacteria found in the soil, if it was absorbed through wounds

C They can be at risk of contracting the hepatitis virus, if they allow the wounds to get dirty

D They can be at risk of contracting septicaemia, if the wounds become infected with contaminants

9.52 Workers exposed to sandstone containing 70 to 90% silica are at risk of developing serious health conditions if they absorb the substance through their skin. Is this correct?

A Yes - this substance is very corrosive, and can quickly burn through skin upon contact

B No - this substance has no negative effects, as it cannot be broken down

C Yes - this substance, when in contact with exposed skin, is highly irritant and toxic

D No - this substance creates harmful dust, causing long-term lung damage when inhaled

9.53 You are reviewing the risk assessment for an engineer who is carrying out a soldering task. They are wearing suitable eye protection and an FFP1 dust mask. What is the significant risk posed to this individual in wearing this type of equipment?

A It will not prevent them from inhaling the harmful toxic fumes generated when carrying out this task

B It will fully protect them against inhaling the toxic fumes and against all potential health risks from this task

C It is not compatible with the eye protection worn for this task, reducing protection against hazardous fumes

D It can deteriorate very rapidly if any hot fumes come into contact with it and can burn through to the worker

9.54 Painters and decorators working on old lead paintwork are at risk of toxic substances entering their body by which of the following two ways?

1	Injection through the skin by the tiny sharp particles created during sanding	A	1 and 2 only
2	Absorption through the skin by directly touching it with fingertips	B	2 and 3 only
3	Ingestion if it is on the hands during eating, drinking or smoking	C	3 and 4 only
4	Inhalation of the fine dust particles created by the sanding process	D	1 and 4 only

9.55 Which **two** of the following are not stated in a control of substances hazardous to health (COSHH) risk assessment?

A Properties of the substance

B Local toxic waste disposal sites

C Activities that create high exposure

D Meanings of symbols on the substance

E Information on health effects

9.56 What procedure must employers put in place if a worker is liable to be exposed to substances hazardous to health?

A Appropriate medical insurance

B Access to a health physician

C Appropriate health surveillance

D Access to mental health support

9.57 What is the main purpose of carrying out health surveillance on employees who regularly work with hazardous substances?

A To detect any medical problems as early as possible

B To allow appropriate first-aid equipment to be in place

C To identify changes needed to substance suppliers

D To protect against future medical compensation claims

9.58 The painters on site are using two-pack epoxy paint. Which significant procedure must their employer ensure is in place?

A A job rotation system to minimise and control exposure

B A training programme to ensure safe and proper use of the product

C A first-aid course addressing what to do if any product touches their skin

D A health surveillance programme to detect and record any health effects

9.60 In a joinery workshop that assembles large quantities of hardwood doors, suitable control measures have been implemented following a control of substances hazardous to health (COSHH) assessment. Which one of the following would be the next stage?

A To update the assessment within the next six months

B To carry out regular monitoring routines to ensure efficiency

C To inform all first-aiders of the potential health risks identified

D To issue the assessment to all affected operatives

9.59 A control of substances hazardous to health assessment has identified that substances used on site are a significant hazard. Which two of the following must be put in place for the health surveillance of all exposed workers?

1 Retention of heath records for a minimum of 25 years from the date of first entry

2 Retention of health records for a minimum of 40 years from the date of last entry

3 Recording of any medical decisions or restrictions by an occupational health professional

4 Recording of confidential medical information provided by a general practitioner

A 1 and 2 only

B 2 and 3 only

C 3 and 4 only

D 1 and 4 only

9.61 You are co-ordinating the clean-up process of a hazardous substance spillage on site. The correct personnel protective equipment (PPE) has been issued to the workers involved - it is important that these workers also maintain a high standard of personal hygiene whilst carrying out this work. Is this correct?

A Yes - it ensures that the PPE issued is kept clean and maintained properly, and can be safely used by other workers after the completion of the task

B No - employers cannot insist that employees keep themselves clean whilst at work, and the PPE covers them up whist completing the task

C Yes - it ensures that cross-contamination is significantly reduced after the PPE is removed, supporting control measures for the task

D No - this has no impact on the quality of work that employees produce, as long as they wear the PPE correctly when completing the task

9.62 A team of labourers need to undertake work activities within a segregated area of the site, which will require the use of hazardous substances. What must now be implemented to ensure that these hazardous working activities are effectively controlled?

A Constant supervision during the tasks, to ensure that workers are adhering to the safe system of work

B Toolbox talks, to make workers aware of how to identify a toxic material and how to use it safely

C Training sessions, to provide workers with information on how to act safely in the event of a toxic spillage

D Instruction and training, to raise worker awareness of the dangers of exposure and the precautions to take

9.63 Which one of the following details must be included in a safe system of work before using a hazardous substance?

A Information on required training

B Information on health effects

C Information on environmental impacts

D Information on the supplier

9.64 Under the Control of Substances Hazardous to Health Regulations (COSHH) 2002, which one of the following legal requirements is an employer not expected to do?

A Identify harmful substances that may be present in the workplace

B Establish how workers will be exposed to the materials and be harmed

C Provide monitoring and health surveillance when appropriate

D Facilitate training sessions on the risks of exposure to lead

9.65 How often must a control of substances hazardous to health (COSHH) risk assessment be updated?

A Regularly to take into account any changes in the workplace

B At the same time as all the other site risk assessments

C Annually on the date of first purchase of the substance

D When new stock of the same substance has been purchased

9.66 The main purpose of using the global harmonised system for the classification and labelling of chemicals is to provide...

A as much information as possible using as many pictograms as necessary

B clear information for the protection of humans and the environment

C easy identification of toxic substances for people from diverse backgrounds

D less written information on the contents and effects of the substances

9.67 A quantity of new hazardous substances have arrived on site and they have been stored together without referencing the safety data sheets. Which one of the following could be a consequence?

A They could get mixed up with old stock and affect application

B They could react with each other and cause a major incident

C They could be logged incorrectly and alter stock control records

D They could be unsafely stacked and overloaded resulting in spillage

9.68 You are writing a control of substances hazardous to health (COSHH) risk assessment. How could the safety data sheet that accompanies the substance help you?

A By confirming how much material can be effectively applied per square metre to complete the task

B By determining the type and correct compatibility of the personal protective equipment required

C By identifying the possible health implications and severity of risk posed to those using the material

D By establishing that the quantities purchased are sufficient in number for completing the task

9.69 During the early stages of site set-up, which of the following must be considered first regarding the storage of hazardous substances?

1	The location of the storage containers	A	1 and 2 only
2	The segregation requirements	B	2 and 3 only
3	The access and egress for delivery	C	3 and 4 only
4	The supply level requirements	D	1 and 4 only

9.70 A large quantity of hazardous substances have been delivered to site. Is it acceptable to store them all together before a control of substances hazardous to health (COSHH) assessment has been completed?

A Yes - but only for a duration that does not exceed 48 hours from their delivery

B No - the correct chemical inventory data must be checked for segregation requirements

C Yes - as long as material safety data sheets have been obtained for each substance

D No - dedicated site storage areas will have lists of substances that can be kept there

9.71 When conducting a health and safety inspection on site, you have concerns about the incorrect storage and segregation of a sub-contractor's hazardous materials. Which of the following **three** are significant factors that have led to this issue?

A Storing substances in the wrong type of area

B Keeping inconsistent records about the substances

C Estimating the wrong quantities required

D Using inaccurate identification labels on containers

E Purchasing the incorrect substances

F Issuing too many keys for the storage area

9.72 You are in charge of setting up a new construction site. What must be done when a Control of Substances Hazardous to Health risk assessment has determined that certain substances on this site will need to be stored separately?

A Create a suitable storage area, with adequate ventilation and extraction facilities, to hold all identified materials within the zone

B Install a suitable storage area, with boxes to hold appropriate safety equipment for material use stored within the zone

C Identify a suitable storage area, and check the materials' compatibility with others to be stocked within the zone

D Make sure a suitable storage area is positioned next to the site office, and that segregation shelves are installed within the zone

OCCUPATIONAL HEALTH, WELLBEING AND WELFARE

10 Manual handling

10.01 A groundworker has been carrying cement blocks from one area of the site to another, and is now complaining of mild back pain. What is the first action you should take, as their manager, to reduce any further strain on the worker's back?

A. Arrange for the blocks to be dropped off nearer where they are required on site

B. Provide the worker with thicker gloves and a back belt for support

C. Ask another worker to help so that the task can be shared

D. Arrange for the cement blocks to be exchanged for blocks made of a lighter material

10.02 Which two of the following would be priorities before carrying out any manual handling lifting activities?

1	Discussions with the workers about the risks involved in the task	A	1 and 2 only
2	A full risk assessment of the task, with the results recorded	B	2 and 3 only
3	Checking and logging the delivery of the load before unloading starts	C	3 and 4 only
4	The installation of extra barriers and signage around the site loading area	D	1 and 4 only

10.03 What do the Manual Handling Operations Regulations 1992 require an employer to do?

A. Manage, control and reduce the risk of injury

B. Allow employees to assess their own abilities

C. Identify and list all heavy loads on site

D. Make employees accountable for their own operations

10.04 From the list below, select the **four** main factors that need to be considered when assessing a manual handling operation.

A Individual

B Environment

C Time

D Load

E Equipment

F Task

G Length

10.05 A worker has come to you with concerns about the load they have been asked to move. They think it might be too heavy for them; they cannot divide it into smaller parts; and there is no one to help them. What would you advise them to do?

A Do nothing until a safe method is identified

B Instruct the forklift operator to help them lift the load

C Try to use the correct lifting method

D Lift and move the load quickly

10.06 In a manual handling risk assessment, which **three** factors need to be considered when assessing each worker's suitability to carry out the task?

A Individual's physical suitability

B Capacity of understanding

C Person's training and knowledge

D Individual's weight

E Health records and previous injuries

F Individual's height

10.07 If your workers are not following the safe system of work when moving slate tiles onto a roof, what must you do?

A Find out why they are doing it like that and change the risk assessment accordingly

B Identify who started doing it that way and issue them with an on-the-spot verbal warning

C Let them carry on, as no injuries have occurred and they are getting through the work faster

D Tell them to stop and that they need to follow the agreed procedure for the task

10.08 Identify **four** psychosocial factors that need to be considered when carrying out a risk assessment of a manual handling task.

A High workloads

B Sufficient management

C Short deadlines

D Poor communication

E Difficult dynamics

F Repetitive tasks

10.09 Which two of the following are needed before an individual lifts a heavy load?

1	Appropriate training and guidance	A	3 and 4 only
2	Personal protective equipment	B	1 and 2 only
3	A back-support belt	C	2 and 3 only
4	A completed risk assessment	D	1 and 4 only

10.10 A surveyor has started complaining of back pain, blaming the regular moving and lifting of the survey equipment in and out of their vehicle. Is this classed as a musculoskeletal disorder injury under the Manual Handling Operations Regulations 1992?

A No, because they are using their own vehicle not related to work until they are on site

B Yes, because they are using the equipment being moved to carry out a work-related task

C No, because they have not mentioned it before and have been doing this role for a while

D Yes, because this is the movement of hazardous equipment to complete a work task

10.11 Under the Manual Handling Operations Regulations 1992, it is an employer's responsibility to do which one of the following?

A Select the correct tools for the task

B Advise how long the task will take

C Implement appropriate control measures

D Inform the worker how much they can lift

10.12 Under the Manual Handling Operations Regulations1992, an employee has a responsibility to do which one of the following?

A Produce a full written risk assessment for the lifting task they are going to carry out

B Follow the manufacturer's instructions when issued with new lifting equipment for the task

C Make any suitable changes to the safe system of work as they carry out the lifting task

D Inform their employer of any shortcomings in the safe system of work provided for the task

10.13 Identify two duties/responsibilities workers have under the Manual Handling Operations Regulations (1992)

1	Follow safe systems of work and revise these as and when required whilst carrying out the task	A	1 and 3 only
2	Inform their employer of any changes they identify or hazards when carrying out manual handling activities	B	2 and 4 only
3	Follow the safe system of work put in place for the task and to co-operate with the employer on all health and safety matters	C	2 and 3 only
4	Identify their own training needs to suitably carry out the tasks and to avoid risks	D	3 and 4 only

10.14 In line with the Manual Handling Operations Regulations 1992, which **three** factors should be in place to protect employees from injury when planning lifting activities?

A Employ the correct people for the work

B Provide appropriate training and information

C Consider everyone's physical suitability

D Supply correct footwear and clothing for the work

E Ensure lifting tasks are shared out evenly

10.15 Do the Manual Handling Operations Regulations 1992 also include the use of mechanical lifting equipment to help with a task?

A Yes, as the mechanical aid does not fully eliminate the task

B No, because the mechanical device is lifting the load

C Yes, because the mechanical aid is doing all the lifting

D No, because the lifting is mechanical, not manual

10.16 Identify **three** risk assessment tools developed by the Health and Safety Executive to help employers comply with the Manual Handling Operations Regulations 1992.

A Risk assessment of pulling and pushing (RAPP)

B Safe system of lifting and moving charts (SSLMC)

C Mechanical lifting aids for manual operation tasks (MLAMOT)

D Manual handling assessment charts (MAC)

E Assessment of repetitive tasks (ART)

10.17 Identify which **two** of the following would not be classed as a manual handling activity under the Manual Handling Regulation 1992 (as amended 2002).

A Working on a scaffold from a height, throwing bricks down a rubble shoot

B Moving a small load by hand, above shoulder height to another area at the same height

C Carrying out maintenance on plant by using a spanner with manual downwards force to release a bolt

D Using a hoist to lift roofing materials to height, then unloading these by hand

E Carrying out demolition of a wall by striking it with force using a large sledge hammer

10.18 From the options below, identify **three** key pieces of information that must be included in a manual handling risk assessment when the activity cannot be avoided, to enable the task to be completed safely.

A Contents of the load

B Full weight of the load

C Where the centre of gravity is

D Material the load is made from

E Where the heaviest side of the load is

10.19 When carrying out a risk assessment for a manual handling activity under the Manual Handling Operations Regulation 1992, which **three** of the following must you take into account when assessing an employee's suitability for the task?

A Whether they are, or have recently been, pregnant

B A disability that may affect their physical capabilities

C Can they work with other colleagues as a team?

D Do they have a medical history of knee or back pain?

E Previous knowledge of the load type to be lifted

F Whether they will need assistance to read the risk assessment

10.20 Why must employees primarily follow the safe system of work provided by their employer for all manual handling activities?

A To avoid causing any accidents or near misses

B To comply with their own legal obligations

C To comply with their employers risk assessment

D To ensure they do not damage the materials

10.21 Which one of the following is not commonly linked to work-related upper limb disorders (WRULDs)?

A Working with hand-held power tools for long periods of time

B Working outdoors in extreme heat with little shade

C Uncomfortable or awkward working posture

D Repetitive work, using the same hand or arm action

10.22 If gloves are mandatory on a construction site, which type of injury will be reduced?

A Cuts and abrasions

B Crushing limbs

C Injection hazards

D Nips and traps

10.23 What causes over half of all work-related ill health among construction workers?

A Occupational cancer

B Work-related stress

C Occupational asthma

D Musculoskeletal disorders

10.24 How can employers address the impact of musculoskeletal disorders on their employees?

A Encourage them to take absence days to recover

B Encourage them to report any pain or discomfort

C Encourage them to take medication to manage pain

D Encourage them to read manual handling guidance

10.27 How might an individual sustain a limb crush injury when carrying out a manual handling activity?

A Moving loads erratically resulting in falling against an object due to loss of balance

B Poor lifting techniques resulting in being trapped by the load

C Wearing poorly fitting clothing that becomes trapped by the load

D Moving materials while wearing the incorrect personal protective equipment

10.25 Why is it important that workers wear appropriate protective gloves when handling cement bricks? Select **two** correct answers.

A They reduce the risk of vibration white finger

B They reduce the risk of cuts and abrasions

C They reduce the risk of skin disease

D They reduce the risk of osteoarthritis

E They reduce the risk of repetitive strains

10.26 By using correct positioning of hands and feet during a manual handling task, limb-crushing injuries would be avoided. Is this correct?

A No, because crushing would still be likely when carrying out the task

B Yes, the likelihood of crushing is reduced through safe manual handling techniques

C No, the injury would not occur if the correct personal protective equipment is worn

D Yes, because the operative has read the risk assessment and method statement

10.28 Which of the following **two** are not common injuries sustained by manual handling activities?

A Knee bursitis

B Lower limb disorder

C Carpal tunnel syndrome

D Musculoskeletal disorder

E Migraines

F Tinnitus

10.29 Which one of the following is the most common cause of a meniscal lesion or tear damage?

A Twisting or bending a knee when carrying a load

B Lifting a small light load too near to the body

C Lifting a large heavy load with a pallet truck

D Twisting the shoulders when loading a cement mixer

10.30 Which one of the following is a collective manual handling solution?

A Wheelbarrow

B Lifting hook

C Rubble chutes

D Keg truck

10.31 Before carrying out a team lift of a heavy load, which one of the following is an important element to consider?

A There are at least four people helping with the load at the same time

B The people lifting the load are about the same size and build

C The people lifting the load are all the same age and gender

D The lifting is carried out on a level concrete surface

10.32 Glaziers on site are installing large sheets of glass into window frames on a new build property. Which type of collective manual handling equipment is most appropriate for this task?

A Suction pads

B Lift hoist

C Lifting hooks

D Mobile conveyor

10.33 Workers are using lifting equipment that is powered by a vacuum pump. What else should be fitted to this equipment?

A An audible and visible warning device

B A safety net situated below the lift

C Replaceable suction cap ends

D Safety chains fitted to the lifting cable

10.34 Your employer has installed gravity rollers to aid the movement of materials around the workshop. This will eliminate all manual handling tasks. Is this true?

A Yes, using this mechanical method removes all the need for a worker to lift the materials

B No, there will still be a need to carry out a small lift of the materials on to the top of the roller

C Yes, using this mechanical method requires no physical involvement from a person whatsoever

D No, this type of machinery still requires a person to lift and hold the materials as they go around

10.35 What is the first thing that an employer needs to consider when introducing vacuum lifts to reduce manual handling tasks?

A The increased costs involved in installing and maintaining this equipment

B The introduction of new risks that will have to be considered and assessed

C The extra training that will be required for the plant mechanics to install

D The new procedures and operating techniques that will be required for safe use

10.36 An employer has introduced automated manual handling aids into their safe systems of work. Which **two** of the following would not be appropriate?

A Battery powered pallet truck

B Cylinder trolley

C Conveyor system

D Roller track

E Powered hoist

10.37 For roofers who are currently transporting tiles up scaffolding by hand, the introduction of which mechanical aid would be most effective at reducing the manual handling risks they face?

A Rotary table

B Dual lifting hook

C Material hoist

D Lift truck

10.38 When manual handling tasks are unavoidable, which **two** solutions could help to significantly reduce the risk of injury?

A Computerisation

B Motorisation

C Standardisation

D Automation

E Mechanisation

10.39 A labourer is experiencing back pain, possibly caused by moving 25kg sacks of materials into a mixer. Which one of the following could stop this injury from happening again?

A Use bigger bags handled by a forklift truck and redesign the feed chutes

B Use smaller bags and rotate the work force and limit working time

C Change the materials used to a lighter product that will be easier to lift

D Stop the work altogether and ask for a design change to the building

CONTENTS

General safety

11 Safety signage and symbols

11.01 Safety signs have different designs and colours to highlight their meaning. What are the different colours of site signage?

- A Red, orange, green and black
- B Blue, orange, yellow and green
- C Red, yellow, green and blue
- D Green, red, yellow, and orange

11.02 Which of the following types of coloured sign correctly indicates a Safe condition or relative place of safety.

- A Black pictogram on a white background
- B White pictogram on a blue background
- C Black pictogram on a yellow background
- D White pictogram on a green background

11.03 What is the purpose of signs that display a round shape with a white pictogram on a blue background?

- A They warn people of a hazard or danger
- B They prescribe certain behaviours
- C They prohibit certain behaviours
- D They inform people what not to do

11.04 Which one of the following types of sign should be used to indicate the location of safety equipment on site?

- A Rectangular or square shape; white pictogram on a red background
- B Rectangular or square shape; white pictogram on a green background
- C Rectangular or square shape; white pictogram on a blue background
- D Rectangular or square shape; white pictogram on a yellow background

11.05 Why is it necessary to place prohibition signs in places where hazards exist?

- A They prevent behaviour likely to increase or cause danger by informing people of things they must not do
- B They warn people of specific hazards likely to increase or cause danger by informing people of things they must do
- C They inform all persons entering the works areas of hazards likely to reduce danger by informing people of actions they should take
- D They prescribe specific behaviours likely to cause danger by informing people of no go areas and equipment to be used

11.06 Safety signs should be used to re-inforce the information you have already covered in a site induction and for workers to use to highlight hazards as their working environments change. Is this correct?

A Yes. Signs must be displayed at the site entrance and should include a selection of mandatory, warning, prohibition, traffic, emergency instruction and first-aid signs

B No. Sufficient time is spent on safety signs in the site induction and workers do not need to have this repeated as they are trained and competent workers

C Yes. Signs should be provided where other methods have been properly considered, cannot deal satisfactorily with certain risks and the use of a sign will further reduce that risk

D No. Safety signs are covered in the site induction and workers knowledge of signs and hazards is tested so there is no need for additional signs to provide information

11.07 What information must fire safety signs and notices display?

1	Actions to be taken in the event of a fire	A	1 and 2 only
2	Instructions for extinguishing a fire	B	1 and 3 only
3	How to call the fire and rescue service	C	2 and 4 only
4	The location of the nearest fire alarm call point	D	3 and 4 only

11.08 Which of the following types of safety sign would be used to provide information about radioactive materials or a source of non-ionising radiation on a construction site?

A Round shape, black pictogram on white background, red edging and diagonal line

B Rectangular shape, white pictogram on a red background with black edging

C Square shape, white pictogram on a blue background with diagonal line

D Triangular shape black pictogram on a yellow background with black edging

12 Fire prevention and control

12.01 You are carrying out emergency planning onsite. What must be in place to make sure you are compliant with the Regulatory Reform (Fire Safety) Order 2005?
Select **two** options.

A A fire risk assessment that is updated on an monthly basis

B A fully revised hot-works risk assessment and permit

C A relevant person to monitor effective evacuation procedures

D A fire risk assessment that is updated to reflect site changes

E An appointed responsible person who controls the premises

12.02 What controls must be in place when work is taking place in a corridor adjacent to an emergency escape route?

A Tools, equipment and materials do not block the route

B All tools and equipment are stored outside the exit doors

C Fire detection is switched off when working in that area

D Only spark proof tools are used within this area

12.03 What must be in place along with the construction phase plan before work starts on a new site?

A Fire risk management plan

B Firefighting emergency plan

C Fire risk assessment

D Fire safety and evacuation plan

12.04 If a fire broke out in a diesel tank, what class of fire would it be?

A Class A

B Class B

C Class C

D Class D

12.05 When positioning a site office for a timber-frame project, it is important that it is situated at least 20m away from the construction because...

A Lorries delivering roof trusses need a larger clearance

B It allows greater access for emergency and rescue teams

C As a high-risk project, the distance creates a fire break

D It gives larger plant onsite more clearance for safe movement

12.06 Under the Regulatory Reform (Fire Safety) Order 2005 what is your main responsibility as the responsible person on site?

A Ensure a suitable fire risk assessment is carried out and reviewed when necessary

B Extinguish small fires that occur on site using fire extinguishers and to record them

C Ensure that the construction phase of the project meets current fire regulations

D Carry out all maintenance and repairs to all firefighting equipment stored on site

12.07 Under the Regulatory Reform (Fire Safety) Order 2005, which two must be priority on a construction site?

1	A fire risk assessment that is regularly reviewed	A	1 and 3 only
2	Sufficient and maintained fire-fighting equipment	B	2 and 3 only
3	An appointed responsible person, either an employer or owner	C	3 and 4 only
4	Suitable and compliant fire escape doors and routes	D	1 and 4 only

12.08 During the design phase of a large construction project, it is the duty of the principal designer or designer to contact the fire rescue service and discuss access requirements and arrangements before any work starts. Is this correct?

A No, this is not a requirement at this stage, this would happen after work had started on the project

B Yes, it is important that all emergency services are informed of any construction projects being undertaken

C No, this responsibility falls upon the appointed person after the fire risk assessment has been completed

D Yes, it is the responsibility of these duty holders and they must update them as the project progresses

GENERAL SAFETY

12.09 There is a legal requirement for the responsible person on site to ensure that a specific fire risk assessment is carried out. This can only be done by this individual. Is this correct?

A Yes, this can only be carried out by the responsible person identified before site work begins.

B No, they can nominate another trained and competent person to do it on their behalf.

C Yes, as long as they are trained and competent to do so, and been appointed by the main contractor.

D No, under no circumstance can this role be delegated to another individual on site.

12.10 What is the first step of a fire risk assessment?

A Remove the risk

B Evaluate the risk

C Identify the hazards

D Identify the people at risk

12.11 When completing a fire risk assessment for a new build project it is important to consider which **three** elements?

A The type of construction materials being used

B The carbon footprint of the construction

C The location of the construction

D Additional charging units for electric vehicles

E How close it is to other buildings

12.12 Although basic fire risk assessment principles are the same, there are differences to consider for new builds compared to a refurbishment of an existing building. Which of the following two cover some of these considerations?

1. The building's location and proximity to other buildings
2. The type of construction materials that will be used
3. Suitable training for all workers in the use of fire equipment
4. All equipment on site is battery or electric powered

A. 1 and 4 only
B. 2 and 3 only
C. 3 and 4 only
D. 1 and 2 only

12.13 For a fire risk assessment to be legally compliant, it must be...

A. reliable and comprehensive
B. suitable and reliable
C. suitable and sufficient
D. comprehensive and sufficient

12.14 When carrying out a fire risk assessment on site, which one of the following would not help you evaluate the risk of a fire occurring?

A. Identifying any potential ignition and heat sources on site
B. Checking which workers on site have had fire awareness training
C. Having a list of all the plant, equipment, and fuel being used on site
D. Knowing the source of all the building materials on site

12.15 On large, complex projects it is important to periodically review the fire risk assessment. Identify one of the following where you do not need to review it?

A. When the project progresses and changes, affecting the evacuation procedures
B. When there are more people on site and have a longer distance to travel to safety
C. When site conditions change as the project progresses, altering the fire risks
D. When the main contractor has adjusted schedules to fit around delivery of materials

12.16 A joinery apprentice with a hearing impairment has started on site. In relation to fire safety controls, what must be actioned?

A Fire extinguisher training is given to the individual as soon as possible

B A site induction is given to highlight the fire controls and processes in place

C The fire risk assessment is reviewed to account for the apprentice's disability and age

D The fire risk assessment is given to the apprentice to read, sign and return

12.17 Which one of the following is not a safe procedure when using a fire blanket to extinguish a small fire?

A The user wraps their hands around the blanket corners

B The gas and electric supplies are switched off

C The user looks under the blanket to check the fire is out

D The blanket is not thrown, but gently placed over the fire

12.18 Which one of the following is not a requirement for fire extinguishers on construction sites?

A They are located at identifiable fire points

B They are appropriate for the risks on site

C They are competently serviced and maintained

D They are replaced every 12 months

12.19 A fire has broken out in a confined space. What would be the primary risk if you used a carbon dioxide extinguisher to put it out?

A It would cause a chemical reaction

B It would reduce visibility significantly

C It would have no effect on the fire

D It would reduce the percentage of oxygen

12.20 A firefighting hose reel is in-situ on a refurbishment project. What is the recommended safe coverage area of such a hose?

A 400 m^2

B 600 m^2

C 800 m^2

D 1000 m^2

12.21 You are working on a re-development project where the water supply to existing firefighting hose reels has been drained. Is this acceptable?

A No, the supply should be made live as soon as possible in progressive stages as the fire load of the project increases

B Yes, as long as the drainage was carried out by a trained and qualified water mains contractor and the hose reels removed

C No, the supply should never be cut off and there should always be a water supply for fire fighting prevention

D Yes, but only if there are a substantial number of water fire extinguishers made available around the project

12.22 An engineer is completing maintenance checks on fire extinguishers. They have to follow a recognised standard. Which **three** rules are correct?

A All extinguishers with a plastic body manufactured after the year 2002 must be replaced as soon as practicably possible

B There must be an appropriate scheme in place to ensure that all extinguishers on site are regularly checked and maintained

C The engineer must ask to see a copy of the fire risk assessment before beginning to service the extinguishers

D The responsible person must ensure that a fire logbook is kept to record all events involving extinguishers

E The maintenance and checking of the extinguishers must be carried out and recorded by the appointed fire warden

12.23 For workers who have been chosen to operate fire extinguishers in case of emergencies, it is essential that they are...

1	suitably trained in their correct use	A	1 and 2 only
2	capable of lifting the extinguishers	B	3 and 4 only
3	over the age of 25 and physically fit	C	2 and 3 only
4	able to identify the correct extinguisher for the fire	D	1 and 4 only

12.24 How does a dry powder fire extinguisher extinguish a class B fire?

A It smothers the whole fire and starves it of oxygen

B It forms a barrier of fine powder against other sources of ignition

C It slowly cools the fire with the fine crystal powder elements

D It creates a dust cloud that absorbs heat and settles over the fire

12.25 On a complex construction site, it is good practice for the 'responsible person' to appoint which other fire safety-related role?

A Fire safety supervisor

B Fire safety co-ordinator

C Fire safety manager

D Fire safety director

12.27 It is deemed good practice to appoint five additional fire safety wardens on complex construction projects. Is this correct?

A Yes, these are the requirements set under the Regulatory Reform (Fire Safety) Order 2005 and must be complied with

B No, there is no specific number of fire wardens required; it depends on the layout and the number of people working on the site

C Yes, these are the requirements set under the Construction (Design and Management) Regulations 2015 for sites and must be complied with

D No, as this would over-complicate the fire safety and evacuation plan prepared for the site

12.26 Which two of the following are responsibilities of an appointed site fire safety co-ordinator?

1 To deploy a fire extinguisher without any previous training

2 To assist in the implementation of the fire safety and evacuation plan

3 To ensure they train all fire wardens in their duties

4 To ensure hot-work permits are followed and managed correctly

A 1 and 2 only

B 2 and 3 only

C 1 and 4 only

D 2 and 4 only

12.28 When fire safety coordinator's are carrying out their duties, which **three** areas must they do to be legally compliant?

A Liaise with the fire and rescue services in the event of a fire

B Carry out a daily role call to account for all workers on site

C Keep records of all inspections, tests, and fire drill practices

D Ensure all fire extinguishers on site are tried and tested weekly

E Carry out weekly inspections of escape route and fire alarms

12.29 You are in charge of setting up a large construction project and you are developing emergency procedures for the site. Which of the following would it be advisable to invite to the site to support this?

A Health and safety executive (HSE) to check and ensure regulations are followed

B Local authority to conduct environmental surveys and identify control measures

C Fire and rescue service to identify water sources and access requirements

D Local Police to advise on current crime trends and appropriate security measures

12.30 How can a fire safety co-ordinator ensure that all site personnel know what to do in the event of a fire?

A Have random conversations with individuals

B Conduct regular fire drill practices

C Distribute leaflets indicating fire procedures

D Conduct all fire-related site inductions

12.31 There has been a major change to the site layout. What must the fire safety co-ordinator review?

A Fire extinguisher numbers

B Emergency sign positions

C Fire evacuation plan

D Emergency mobile phone coverage

12.32 In the event of a fire on site, how can a fire safety co-ordinator ensure the safe evacuation of visitors?

A Discourage site visitors, but if they must visit, ensure 'visitor safe areas' are established

B Have arrangements in place to log visitors in and out, so that they can be located quickly

C Issue all visitors with high-visibility clothing and visitor badges so that they are easily identified

D Create a visitor buddy system, so there is always someone with them who knows the fire procedure

12.33 A hot-work permit has been issued for a large timber-frame project. Each day that the permit is in force, what is the minimum amount of time required between the hot works ending and the site closing?

A Thirty minutes

B Two hours

C One hour

D Ninety minutes

12.34 When undertaking a hot works activity in a confined space, what must be completed prior to the work commencing?

A Thorough ventilation for 48 hours to remove any flammable or explosive gases

B Additional temporary ventilation installed and a fire extinguisher made available

C An atmospheric test that confirms there are no flammable or explosive gases present

D The sub-contractors conducting the work have reviewed their own 'permits to work'

12.35 A roofer is carrying out a task in line with a hot works permit. What must they be able to do?

A Complete the work within the timescale

B Administer first aid treatment if required

C Make adjustments to the fire risk assessment

D Select and use a portable fire extinguisher

12.36 What was introduced by the Fire Protection Association (FPA) to increase the competence of operatives who carry out hot works?

A Certification scheme

B Association membership

C Passport scheme

D Hot-work permits

12.37 Hot-work tasks taking place on a timber-framed building must stop at least two hours before the site closes. What is the main reason for this?

A It allows the site to be monitored for any signs of fire before the area is vacated

B The timber structure can distort in the heat, and it allows time to rectify any deformities

C The workers need enough time to tidy up and lock away the tools and equipment

D If the work activity runs over schedule, it can impact on the legal working hours

12.38 Which **two** of the following would not normally be included in a hot works permit?

A Proposed start time and duration of the hot works

B Instructions for checking for combustible materials in adjacent locations

C Description and location of where the work will be completed

D Information about the porosity content of materials being worked on

E Identification and list of the people carrying out the works

12.39 When hot work activities are taking place on site, the individuals involved must be aware of ...

1	How heat might be transmitted to other areas	A	1 and 3 only
2	How to raise an alarm in the event of a fire	B	2 and 4 only
3	The fire risks in the working location	C	2 and 3 only
4	Other people's capabilities to help in an emergency	D	3 and 4 only

12.40 When carrying out a safety inspection on site, you observe a disc cutter being used. Would a hot works permit be required for its use?

A Yes, they have the potential to cause a fire

B No, sparks would not be generated from its use

C Yes, all tools require a hot-work permit for use

D No, this tool does not generate enough heat

12.41 Which of the following is not a requirement for a compound where liquefied petroleum gas containers are stored on site?

A It is on a level base of compacted earth, concrete or paving slabs

B It is surrounded by a secure chain fence at least 1.8 metres high

C It has sufficient shelter to prevent cylinders being exposed to extreme weather

D It has a base on timber flooring set at a 10% angle to aid water run off

12.42 What does this warning label mean?

A Explosive environment

B Explosive atmosphere

C Explosive substance

D Explosive warning

12.43 When there is no suitable external storage for highly flammable materials on site, it is acceptable to store them within a building. Is this correct?

A Yes, so long as it is in a secure internal fireproof storeroom with fire-detection and suppression systems installed.

B No, under no circumstances can you store highly flammable materials inside, it must be outside in a secure compound.

C Yes, so long as it is in a lockable room, away from the main part of the building with only one person access.

D No, under no circumstance would any highly flammable materials be stored on site, as minimum quantities are used.

12.44 A quantity of gas cylinders, some containing liquid petroleum gas and others oxygen, are being stored together in an external compound.
What is the main significance of this?

A A large volume of highly flammable gas that needs reducing is present on site

B Safe site management and storage procedures of highly flammable gases are being followed

C There is a high fire risk and the oxygen must be stored separately from the other cylinders

D Safe quality checks can be carried out on new cylinders delivered to the storage area

12.45 Explain why liquefied petroleum gas (LPG) cylinders, apart from cylinders used to provide fuel for powered plant, must never be placed on their side during use?

A Because they would give a faulty reading on the contents gauge, resulting in flashback

B Because air could be drawn into the cylinder, creating a dangerous mixture of gases

C Because the liquid gas would be at too low a level to allow the torch to burn correctly

D Because the liquid gas could be drawn from the cylinder, creating a safety hazard

12.46 Sometimes, it is unavoidable to store bulk quantities of highly flammable substances on site. In this circumstance, which two of the following must be applied to the storage area?

1	It is within a pervious bund with a capacity greater than 100% of the total contents stored	A	1 and 2 only
2	It is equipped with heat detectors linked to an automatic sprinkler system	B	2 and 3 only
3	It is within an impervious bund with a capacity greater than 110% of the total contents stored	C	3 and 4 only
4	It is equipped with a quantity of absorbent material to soak up any spilt liquid	D	1 and 3 only

12.47 When using liquefied petroleum gas cylinders on site it is important to follow which **three** procedures?

- A Keep all types of gas cylinder together in the one area
- B Turn off cylinders' valves before connecting any equipment
- C Check cylinders and associated equipment before use
- D Ensure all cylinders are used horizontally to aid release of gas
- E Minimise the amount of gas cylinders on site at one time

12.48 Which one of the following statements is false regarding liquefied petroleum gas (LPG)?

- A A spillage of LPG can create a large vapour cloud of gas that is capable of ignition from some distance away
- B Storage of spare LPG cylinders must be outside the site perimeter in a sealed, secure, non-combustible unit
- C LPG is a colourless, odourless liquid that floats on water but vaporises to form a gas that is heavier than air
- D LPG cylinders and regulators used to heat welfare facilities must be kept outside and be piped in using rigid copper pipes

13 Electrical safety, tools, equipment, lasers and drones

13.01 Is it a legal requirement that portable appliance testing (PAT) must be carried out yearly on all portable electrical equipment?

A) No, the law does not state a frequency, so the employer decides the level of maintenance needed according to the risk

B) Yes, the law states this must be carried out annually by a trained and registered electrician to comply with legislation

C) No, this is not a legal requirement, all portable electrical equipment falls under the responsibility of the user as it is low risk

D) Yes, this is a legal requirement and must be carried out for all electrical equipment including items that are hard wired into the mains

13.02 What is the standard voltage used on sites for all electrically powered equipment?

A) 50 volts

B) 110 volts

C) 230 volts

D) 400 volts

13.03 When electrical tools are being used on site, which one of the following regulations would not need to be complied with?

A) The Personal Protective Equipment at Work Regulations 1992

B) Electricity at Work Regulations 1989

C) Provision and Use of Work Equipment Regulations 1998

D) Management of Health and Safety at Work Regulations 1999

13.04 How must mains power supplies on site be regulated?

A) It must be installed and inspected in line with the Health and Safety at Work etc. Act (1974)

B) It must be maintained with the Electrical Equipment (Safety) Regulations 2016

C) It must be installed, inspected and maintained in line with the Electricity at Work Regulations 1989

D) It must be maintained in line with the Provision and Use of Work Equipment Regulations (1998)

13.05 Your initial site survey has indicated that electricity will be the main power source. Which one of the following would be the safest way of protecting both workers and the main electrical supply?

A Have a residual current device built into the main switchboard to protect the electrical supply

B Make sure that all staff have received suitable training, instruction and supervision

C Provide workers with a residual current device that must be used on all electrical equipment

D Issue all workers with a meter testing kit that they can use to check wattage supplies

13.06 Work needs to be carried out near live electricity cables and a risk assessment has been undertaken. Identify which other document must be in place before work can begin?

A Variation order

B Inspection checklist

C Electrical certificate

D Permit to work

13.07 LOTOTO (Lockout, Tagout, Tryout) is now recognised as best practice. Identify which **three** of the following work procedures would require this process to be applied?

A Changing the blade on a table saw

B Replacing cartridges on a nail gun

C Connecting temporary lighting

D Replacing a belt on a generator

E General maintenance on machinery

13.08 Refurbishment projects can present a risk to workers being electrocuted from live circuits. Which two of the following procedures would mitigate this risk?

1 Installing warning signs and barriers advising of live electrical systems

2 Isolation of the electrical system where work is being carried out

3 Use of suitable tools and equipment, either battery operated or 110 volt

4 Workers complete and follow their own risk assessment for the tasks

A 1 and 4 only

B 2 and 3 only

C 1 and 3 only

D 2 and 4 only

13.09 A competent person must make sure that the power source to a conveyor is isolated before maintenance work can begin. Which **three** procedures must the competent person follow?

A Use an electrical test unit to confirm that the circuit supplying the conveyor is dead

B Turn the power supply off and tape down the switch with warning tape

C Lock off the switch that operates the conveyor so that it cannot be turned back on

D Unplug the conveyor power connection and remove the fuse

E Make sure that the equipment and methods used to check the voltage are reliable

13.10 Who must retain the key when a permit to work involves the use of a padlock to lockout electrical isolators?

A The power supply company's engineer

B The main electrical contractor's supervisor

C The person who is operating the machine

D The person identified on the paperwork for the task

13.11 The power supply to a machine has been isolated to allow maintenance work to be conducted safely. Which two of the following will ensure the power cannot be switched back on again without the correct authority?

1	The person carrying out the work is in control of the main power switch until work is completed	A	1 and 4 only
2	The key to release the lockout tag is held by the person carrying out the work at all times	B	2 and 3 only
3	The fuse supplying the machine has been removed by the person carrying out the work	C	1 and 2 only
4	There are suitable notices and signs in place stating the power must not be switched on	D	3 and 4 only

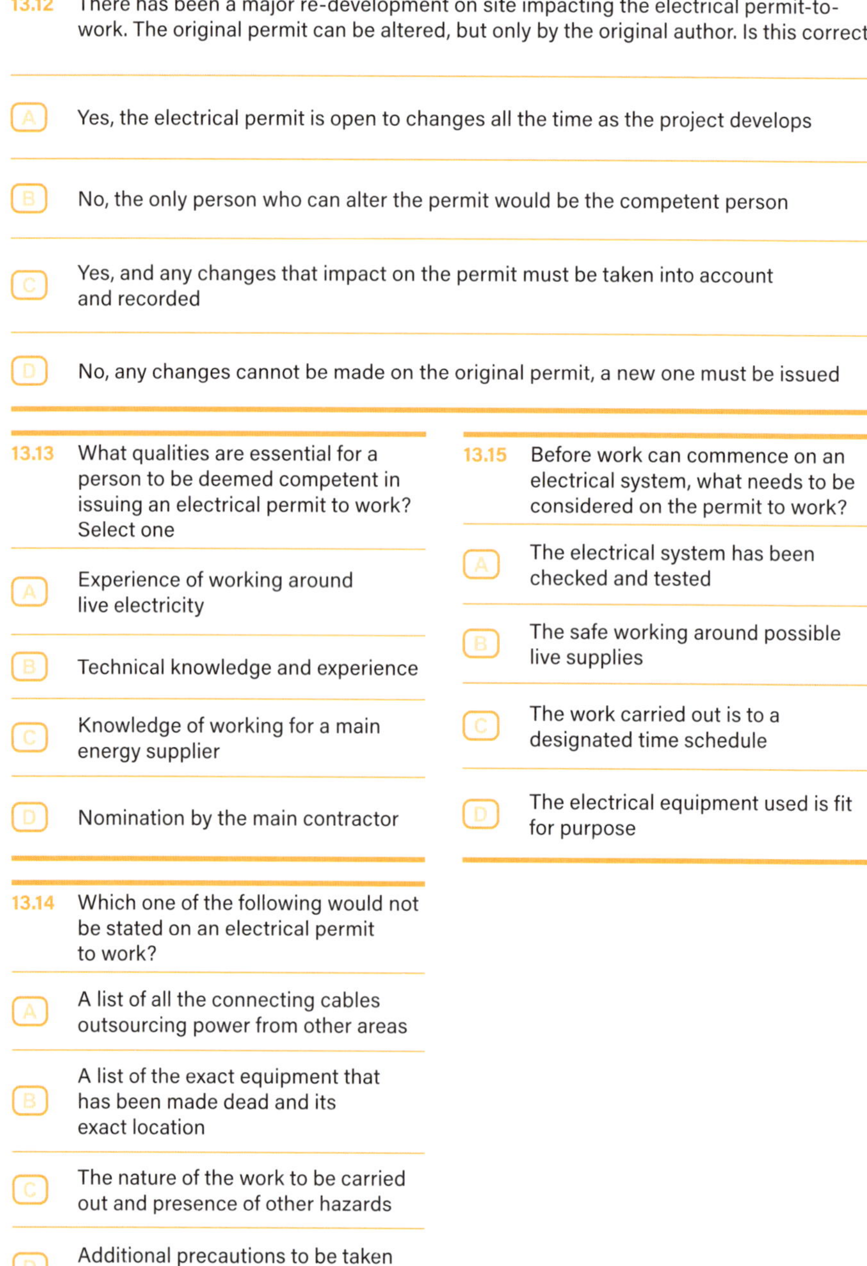

13.12 There has been a major re-development on site impacting the electrical permit-to-work. The original permit can be altered, but only by the original author. Is this correct?

A Yes, the electrical permit is open to changes all the time as the project develops

B No, the only person who can alter the permit would be the competent person

C Yes, and any changes that impact on the permit must be taken into account and recorded

D No, any changes cannot be made on the original permit, a new one must be issued

13.13 What qualities are essential for a person to be deemed competent in issuing an electrical permit to work? Select one

A Experience of working around live electricity

B Technical knowledge and experience

C Knowledge of working for a main energy supplier

D Nomination by the main contractor

13.14 Which one of the following would not be stated on an electrical permit to work?

A A list of all the connecting cables outsourcing power from other areas

B A list of the exact equipment that has been made dead and its exact location

C The nature of the work to be carried out and presence of other hazards

D Additional precautions to be taken during the course of the work

13.15 Before work can commence on an electrical system, what needs to be considered on the permit to work?

A The electrical system has been checked and tested

B The safe working around possible live supplies

C The work carried out is to a designated time schedule

D The electrical equipment used is fit for purpose

13.16 What must be followed if a permit remains in place after a shift changeover? Select one

A The permit is endorsed by the authorised person and transferred to the second operative

B The permit is amended to reflect the changes and endorsed by the second operative

C The permit is displayed in the working area for the second operative to refer to

D The permit is handed over to the second operative so that they can continue the work

13.17 What is the main feature of a direct-acting cartridge operated tool?

A They are heavier to hold

B They have a short battery life

C They tend to be high powered

D They are lighter in weight

13.18 Who should be contacted about the disposal of unused or unserviceable cartridges?

A The contractor

B The manager

C The supplier

D The operator

13.19 Which **three** of the following must be compatible for the safe use of gas-operated fixing tools?

A Power voltage

B Type of cartridge

C Base material

D Type of fitting

E Operative's age

13.20 Which of the following must be performed before firing a first fixing with a cartridge tool?

A Driving a fixing of the intended type into the base material with a hammer

B Issuing the risk assessment and method statement for use of fixings

C Explaining to the operatives the safe use of fixings through a toolbox talk

D A thorough examination of the tool and the intended fixings

13.21 Fully dispensed cartridges are classed as hazardous waste and must be disposed of using the correct procedures. Is this correct?

A Yes, there are strict guidelines for the correct disposal process

B No, they are not hazardous so normal disposal procedures apply

C Yes, they are toxic so need to be disposed of by a specialist contractor

D No, fully dispensed cartridges can be cleaned and re-used

13.24 When operating cartridge tools, which one of the following is not a primary factor for their safe use?

A Batteries must be fully charged before use

B Access to the work area is restricted during fixing activities

C Operators are appropriately trained and supervised

D Tools comply with BS 4078-2 tool specification

13.22 When storing cartridges on site for use in a cartridge tool, which two of the following are required?

1 Arrange them in order, according to their strength and colour

2 Ensure sub-contractors follow their own procedures for storage

3 Check with the local authority to determine if they need to be licensed

4 Provide a secure, dry and cool storage area

A 1 and 3 only

B 2 and 4 only

C 1 and 2 only

D 3 and 4 only

13.23 When charging lithium-ion batteries on site, which **three** safe working practices must be followed to reduce the risk of fire?

A Use overheat protection

B Never use a mains supply

C Never over charge or trickle charge

D Use a temporary supply

E Use time-out protection

13.25 Which one of the following classes of laser levels is the safest to use and poses the least risk?

A Class 1

B Class 2

C Class 3

D Class 4

13.26 What is the significant difference between a class 1 and class 3 laser level?

A Operators of class 3 laser levels must be trained and experienced, where as class 1 operators do not

B Class 1 equipment has been banned within the United Kingdom, where class 3 has not

C Operators of class 3 laser level do not need to be trained and competent but operators of class 1 must be

D Class 1 equipment requires operators to wear eye protection, but class 3 does not

13.27 To operate a laser level, you must be which one of the following

A Colour blind

B Competent

C Aged over 18

D Glasses wearer

13.28 Which of the following must be in place when operating static lasers?

A Screens and warning signs

B Supervision and monitoring

C Exclusion zones and warning signs

D Permission from the local authority

13.29 When completing a risk assessment for the safe use of a large drone on site, which **three** controls must be implemented?

A Plant and crane operatives are made aware of drone activity

B Suitable, sufficient landing areas are allocated

C The flight has permission from the Civil Aviation Authority

D It is feasible for the drone to operate safely within the area

E There is suitable and sufficient support personnel on the ground

13.30 What must be completed by the drone pilot before a flight is undertaken on site?

A Equipment is checked and fully charged

B A method is in place for the aerial work

C The camera is not damaged

D The weather conditions are acceptable

13.31 To allow the safe operation of a drone on site, identify which two procedures must be followed.

1	Designate a safe area for the take-off and landing of the drone	A	1 and 2 only
2	Avoid unnecessary distractions or communication with the drone operator during flight	B	2 and 3 only
3	Ensure that site personnel have stopped work activities while the drone is in flight	C	3 and 4 only
4	Direct any high voltage lighting downwards to avoid distraction to the drone systems	D	1 and 3 only

13.32 Specific incidents and accidents involving drone activities on site are reportable to...

- A The Health and Safety Executive
- B The Local Authority
- C The Civil Aviation Authority
- D The Environmental Agency

14 Site transport safety and lifting operations

14.01 Which **two** of the following are required for a individual to act as a signaller when a load is being lifted or lowered?

- A Awareness training of the type of crane in use
- B Supervision by an authorised crane operator
- C Knowledge of the approved method statement
- D Knowledge of all lifting accessories and connectors
- E Training and certification in, and experience of, lifting operations

14.02 If a plant operator loses sight of the banksman during a manoeuvre, what must they do?

- A Not deviate from the planned route
- B Stop the vehicle immediately
- C Use the two-way radio to locate them
- D Only travel forward in the vehicle

14.04 You are managing groundworks for a project that has a number of deep excavations. How would you prevent heavy vehicles from operating too close to the edge of an excavation?

- A Set up and use stop blocks to stop over-running the excavation edge
- B Set up a traffic management system to plan access around allocated time periods
- C Set up an audible warning system that activates when too close to the excavation edge
- D Set up a system of physical barriers a distance of 3 metres from the excavation edge

14.03 You are managing a large office project, and an incident between two vehicles has resulted in a delivery lorry overturning. What could you do to reduce the risk of this type of incident happening again?

- A Implement designated exclusion zones for deliveries and other site vehicles
- B Use banksmen to control the movements of delivery vehicles from now on
- C Erect safety warning signs at vehicle crossing points and around the site
- D Effectively plan traffic routes to control the movements of all vehicles

14.05 You are managing a site where currently only 'muck away' activities are going on. What action do you take to ensure pedestrians and vehicles are adequately separated?

A Set up pedestrian-only areas from which vehicles are excluded at busy times

B Ensure pedestrians are provided with a clear view of all traffic routes on site

C Set up vehicle-only areas especially where space is limited or traffic is heavy

D Ensure pedestrians are not crossing any vehicle routes on site at any time

14.06 When a banksman is giving directions to the lifting equipment operator, which **two** of the following must be complied with?

A The Health and Safety (Safety Signs and Signals) Regulations 1996

B Traffic Signs Manual Chapter 8 road works and temporary situations 2009

C BS7121-1 Safe use of Cranes - General

D The Traffic Signs Regulations and General Directions 2016

E Lifting Operations and Lifting Equipment Regulations 1998

14.08 Which one of the following is a method used by a banksman to direct plant on site?

A High-visibility flags

B Coloured paddles

C Hand signalling

D Tannoy system

14.07 Which one of the following should occur when a delivery driver and banksman are using portable devices to communicate with each other during a vehicle manoeuvre on site?

A The banksman will need to stand in a safe position where they can guide the reversing vehicle and be visible to the driver at all times

B The driver will be permitted to reverse the vehicle if they lose sight of the banksman as long as they are communicating by portable radios

C The banksman will have to stand close to the vehicle when it is reversing to ensure they can be seen by driver at all times in the reversing camera

D The driver will need to take extra precautions for visibility to ensure they keep sight of the banksman in poor or low light conditions

14.09 During site operations, it is not possible for a plant operator to achieve suitable all-round visibility from their operating position. Which **two** of the following measures would be most effective to protect people from harm?

A Ensuring that lights are fitted to plant, and are used to adequately illuminate the work area

B Providing a competent, qualified vehicle marshaller to work exclusively with the plant operator

C Ensuring that amber flashing beacons are fitted to the vehicle, to warn of its presence

D Implementing exclusion zones to prevent unauthorised access into any danger areas

E Providing information on how to safely signal the operator and receive a positive response

14.10 If vehicles have to reverse in areas where other workers cannot be removed, you should ensure that:

1	All pedestrians in the area wear high-visibility clothing	A	1 and 2 only
2	The driver has sufficient direct vision behind the vehicle	B	2 and 3 only
3	A trained signaller is used to control the manoeuvre	C	3 and 4 only
4	Any turning areas are kept free from parked vehicles	D	1 and 4 only

14.11 Who should carry out an assessment of the ability of the ground or surface to accept the loads from cranes or mobile lifting equipment?

A Crane operator

B Lift planner

C Lift supervisor

D Site engineer

14.12 Following the confirmation of all hazards associated with a lifting operation, which one of the following would be identified in the second stage of the risk assessment?

A What type of crane and equipment will be needed to perform the lift

B The correct sequence and method of how to perform the lift safely

C Who might be harmed by the risks and consequences of any harm

D The contacts details of anyone involved in the emergency procedures

14.13 When planning a lifting operation, which one of the following category lists is used to establish the complexity of the lift?

A Critical, intricate and multifaceted

B Basic, intermediate and complex

C Low, medium and high

D Elementary, difficult and focused

14.14 Which one of the following is the correct sequence to follow when planning a safe and effective lifting operation?

A Categorise the lift and select the correct lifting equipment and lifting accessories to lift the load, record the details in a lift plan

B Identify the hazards associated with the task and identify control measures, record the lifting plan with a method statement

C Identify the task to be undertaken and survey the site, identify the hazards associated with the task and categorise the lift

D Survey the site and identify the hazards associated with the task, identify control measures and develop the method to be used

14.15 When carrying out a lifting operation from a lorry loader, which **three** of the following are potential risks for the pedestrian operator who is at ground level?

A Falling on uneven ground

B Having an obstructed view of the lifting equipment

C Misinterpreting the banksman's signals

D Inclement weather conditions

E Being struck by the load

F Being unaware of the lift sequence

14.16 A lifting operation on a small and restricted site footprint is being carried out using a radio communication system. Which of the following may cause potential problems for a safe lift?

1	Interference from other radios	A	1 and 2 only
2	Stopping the lift quickly	B	1 and 3 only
3	Loss of signal	C	2 and 4 only
4	Sunny weather conditions	D	1 and 4 only

14.17 An excavator carrying out a lifting operation on site has overturned but no-one was injured. Which **two** of the following must happen next?

A Interview the site manager and site supervisor

B Carry out a thorough investigation of the incident

C Make a report to the Health and Safety Executive (HSE)

D Report the incident in the company accident book

E Update the lift plan following the incident

14.18 When hiring a crane to carry out a contract lift, the majority of the legal responsibility for its safe operation is transferred to the crane hire company. Is this correct?

A Yes, you will be provided with a lift supervisor, slinger or an appointed person, and ground bearing capacity calculations

B No, you must retain full responsibility and accountability for any lifting operations undertaken by any contractors on site

C Yes, you will be hiring the services of the crane company to organise and carry out the whole lift operation

D No, the crane owner has to provide only a crane that is tested, certified, has suitable lifting accessories and a competent operator

14.19 A crane is being hired to undertake lifting operations on a project. What are the employing organisation's duties?

1	To plan the lift and operate a safe system of work	A	1 and 2 only
2	To provide a crane that is properly maintained	B	2 and 3 only
3	To supply a competent lift planner	C	1 and 3 only
4	To provide a crane that is properly maintained	D	2 and 4 only

14.20 When hiring a contract lift crane for site work, which of the following is the contractor responsible for?

1	The overall load bearing capacity calculations	A	1 and 2 only
2	The dimensions and weight of the loads to be lifted	B	2 and 3 only
3	The planning of the lift and operating a safe system of work	C	1 and 4 only
4	The organisation and control over the lifting operation	D	3 and 4 only

14.21 When establishing the complexity of a lifting operation, the complexity of the load will include characteristics such as the rated capacity (safe working load (SWL)) of the lifting equipment. Is this correct?

A Yes. Load complexity includes the rated capacity (SWL) of the lifting equipment and should be specified by the manufacturer

B No. Load complexity includes characteristics such as weight, centre of gravity and presence of suitable lifting points

C Yes. Load complexity includes the suitability of the ground to take the loads imposed by the lifting equipment during the lift

D No. Load complexity includes the rated capacity (SWL) stated on the current report of thorough examination, issued by the competent person

14.22 Which of the following is the first stage of a lift planning process?

A Carrying out a site survey

B Identifying the associated hazards

C Carrying out a risk assessment

D Identifying the task to be undertaken

14.23 A complex tandem lifting operation with two mobile crawler cranes has been planned for your site. Which one of the following should happen immediately before the lifting operation commences?

A Review the lift plan to check if aspects have changed

B Communicate the lift plan to all persons involved

C Ensure a suitable exclusion zone is put in place

D Ensure all plant has been thoroughly examined

14.24 In line with the Lifting Operations and Lifting Equipment Regulations (1998), how long must the report of a thorough examination be retained after lifting equipment is examined before its first use?

A Until the next report is made

B Two years after the report is made

C One year after the report is made

D Until the user ceases to use the equipment

14.25 Which one of the following practical measures should be considered to ensure that mobile work equipment can be used safely on site?

A Pedestrian routes should be segregated from mobile plant and vehicles, with access points restricted and clearly marked

B Pedestrian routes should be planned in order to minimise disruption to the works on site, congestion and risk of collision

C Pedestrian routes should be established away from any overhead cables, underground cables, sewers, ducts or services

D Pedestrian routes should be segregated from mobile plant and vehicles, either by a safe distance or by physical barriers

14.26 The most effective way to manage the safe passage of pedestrians on a site is to...

A paint pedestrian walkways on all haul roads

B use signs to highlight risks of heavy plant operations

C separate pedestrian access from traffic routes

D restrict pedestrian access to specific times

14.27 Which one of the following actions should be taken to ensure safe vehicle operations and adequate segregation of pedestrians?

A. Place restrictions on pedestrian movements to and from the site

B. Locate materials storage areas away from primary traffic routes

C. Have sufficient room for vehicle movements and safe exit points

D. Provide signallers at crossings to warn drivers and pedestrians

14.28 What is the most effective way of preventing accidents with pedestrians and mobile work equipment on site?

A. Restrictions on pedestrian movements on site

B. Exclusion zones to prevent unauthorised access

C. Amber flashing beacons and reversing alarms on plant

D. Trained vehicle marshals supervising plant movements

14.29 Your company needs to carry out emergency street works involving the use of road-cutting saws and disc cutters in a busy high street. Which **two** of the following are the most effective practical measures to reduce the risk of sparks and stone chips entering pedestrian areas?

A. Shut down areas within a ten metre radius of the work

B. Work at times when the public is less likely to be in the area

C. Provide solid barriers adjacent to public areas

D. Use stacks of materials as temporary noise barriers

E. Provide suitable acoustic curtains near public areas

14.30 When establishing pedestrian routes on a site, they should...

1	provide vehicle and plant operators with a clear view of pedestrian routes	A	1 and 2 only
2	have competent vehicle marshals available at pedestrian crossing points and gates	B	2 and 3 only
3	be wide enough to accommodate the number of people likely to use them at peak times	C	3 and 4 only
4	provide pedestrians with a clear view of traffic movements at crossing points	D	1 and 3 only

14.31 Which one of the following practical measures should be considered when planning traffic and pedestrian management to prevent pedestrians being struck by vehicles or their loads?

A Ensure any traffic route provided is wide enough to accommodate a pedestrian walkway

B Ensure availability of competent vehicle marshals to monitor safety on pedestrian walkways

C Ensure safe, designated, pedestrian routes are available at all work locations and are maintained

D Ensure on-site parking for private vehicles is located as near as possible to the work areas

14.32 Which of the following is the most effective practical traffic and pedestrian management measure that will prevent pedestrians being struck by vehicles or their loads?

A Ensure the site implements clearly sign posted one-way traffic systems with speed limits

B Ensure parking areas are connected to the entrance to avoid visitors crossing site to gain access

C Ensure all workers and visitors on site are issued with suitably coloured high-visibility clothing

D Ensure a car park is provided for privately owned vehicles with direct access to working areas

14.33 The most effective way of managing the risks from reversing vehicles on a construction site is to ensure all vehicles have appropriate reversing warning devices. Is this correct?

A No. Warning systems offer the lowest level of protection and if they are the only precaution to be used, they are only appropriate for low-risk situations

B Yes. Reversing warning lights and alarms provide both visual and audible warnings and are appropriate precautions for use in high-risk situations

C No. Competent marshals or signallers should be in place to warn drivers that they have reached the limit of the safe reversing area

D Yes. Reversing warning lights, alarms and CCTV are fitted as standard as an effective way of managing the risks from reversing

14.34 The majority of construction transport accidents involving pedestrians result from poor operating or driving practices. Is this correct?

A Yes. The majority result from inadequate driver training and inappropriate vehicle use

B No. The majority result from the inadequate segregation of pedestrians and vehicles

C Yes. The majority result from manufacturers' instructions for safe use being disregarded

D No. The majority result from unsafe loading and transportation of materials on vehicles

14.35 To ensure pedestrian safety on construction sites, principal contractors must ensure that pedestrians and vehicles are adequately segregated by establishing...

1	safe loading and transportation of materials on vehicles	A	1 and 2 only
2	pedestrian areas from which vehicles are totally excluded	B	1 and 3 only
3	control of pedestrian movements during construction work	C	3 and 4 only
4	vehicle-only areas especially where space is limited on site	D	2 and 4 only

14.36 Which **two** of the following are the most effective ways of preventing vehicles or their loads coming into contact with people on a construction site?

A Controlling pedestrian movements during construction work

B Providing drivers with a clear view of pedestrian movements

C Pedestrianised areas from which vehicles are totally excluded

D Providing gates for pedestrians at traffic route crossing points

E Establishing safe designated pedestrian routes to work areas

14.37 Which **two** of the following are the most effective ways to prevent pedestrians being struck by vehicles or their loads?

A Providing traffic routes that are wide enough to accommodate a pedestrian walkway

B Presence on site of competent vehicle marshallers for all vehicle reversing movements

C Availability and maintenance of designated pedestrian routes to all work locations

D Provision of onsite parking for private vehicles allowing direct pedestrian access to work areas

E Separation of pedestrian walkways from plant and vehicles by physical barriers

14.38 When planning the layout of a site, which two of the following precautions will reduce accidents related to mobile plant and site logistics?

1	Design groundworks and landscaping to reduce the need to import fill or take spoil off site	A	1 and 2 only
2	Consider ways of moving materials without using vehicles	B	2 and 3 only
3	Design storage compounds to allow vehicles to turn and park	C	1 and 3 only
4	Consider adding extra temporary roadways to the site's permanent road system	D	3 and 4 only

14.39 Which of the following measures can be taken when planning your site layout to minimise the risk of vehicle related accidents?

1	Appoint a trained signaller who can control vehicle movements	A	1 and 2 only
2	Check all vehicles are fitted with appropriate reversing alarms	B	2 and 3 only
3	Provide drive-in/drive-out access to delivery and work areas	C	1 and 4 only
4	Ensure pedestrian and vehicle entry and exits are kept separate	D	3 and 4 only

14.40 When planning site layout to minimise the risk of vehicle-related accidents, which **two** of the following measures must be taken?

A Design traffic routes with turning circles to avoid the need for reversing or similar turning areas

B Consider ways other than vehicles for moving materials to where they will be used

C Ensure trained signallers are available to control all vehicle movements on site

D Ensure drivers can see site exits and pedestrians before crossing the footway

E Design groundworks to reduce the need to import fill or take spoil off site

14.41 The majority of construction accidents involving reversing vehicles can be avoided by careful consideration at the design stage of a project. Is this correct?

A Yes, because there are many accidents and deaths due to a lack of planning and control, especially when reversing vehicles on site

B No, because all workers are competent in all aspects of site vehicle safety, so reversing is no longer a problem on site

C Yes, because so many accidents have been caused by reversing vehicles, all traffic movements must be strictly controlled by a team of experts

D No, because this is explained in detail in the site induction so everyone knows the associated risks and methods to follow

14.42 Which of the following should be taken in to account when assessing the likelihood and potential consequences of a vehicle roll-over on site?

A Measures that can be taken to avoid sudden movements of the mobile work equipment

B Nature of the mobile work equipment and any attachments or accessories fitted to it

C Risk of people being injured if the mobile work equipment comes to a sudden stop

D Warning devices to provide equipment operators with an indication of imminent danger

14.43 Which **two** of the following measures can reduce the risks to the public on partially completed housing development sites?

A Phased occupation of dwellings so that site traffic can be excluded from occupied areas

B Delivery management and preventing unplanned vehicles to enter the site

C Segregated public vehicle and pedestrian routes from site vehicles and site worker routes

D Providing direct access to work areas via established primary pedestrian and traffic routes

E Positioning site offices and welfare facilities a safe distance from primary site traffic routes

14.44 To prevent construction vehicle operations endangering the public, which **two** of the following practical precautions must be taken?

A Suitable restrictions placed on traffic and pedestrian movements

B Appropriate pedestrian and vehicle traffic management systems

C Effective vehicle immobilisation systems and key custody procedures

D Relevant information and signs for visitors on site traffic management

E Suitable turning circles and established one way traffic systems

14.45 Which one of the following is not part of the duties of the site supervisor in relation to mobile work equipment?

A Checking the operator has undertaken the relevant daily checks

B Ensuring all quick hitches are securely locked before starting work

C Ensuring the inspection check sheet has been completed and signed

D Checking the manufacturer's machine operating manual is in the cab

14.46 At the project planning stage, which of the following must be considered for the safe management of deliveries to the site?

A Restrictions on site vehicle use

B Proximity to schools and hospitals

C Drivers working in public areas

D Warning signs for pedestrians

14.47 Which **two** of the following should be considered at the project planning stage for the safe management of deliveries to the site?

A Proximity to schools, hospitals and major shopping centres

B Driver visibility, particularly when working in public areas

C Pedestrians or vulnerable groups passing the site

D Restrictions on site vehicle use in hazardous areas

E Re-routing delivery traffic away from steep hills and bridges

14.48 When planning deliveries to site. Which two of the following will help reduce the risk to pedestrians and drivers of other vehicles?

1	Minimise the number of deliveries to the site during and after busy periods	A	1 and 2 only
2	Consider local hospitals, schools, and site access and exit at busy periods	B	1 and 3 only
3	Divert site traffic away from pedestrianised areas and built-up housing estates	C	2 and 3 only
4	Divert pedestrians from the site to provide adequate space on site for offloading	D	2 and 4 only

14.49 Before a construction project commences, it is important to consider pedestrians and drivers of other vehicles outside of the site. Is this correct?

A No. During the planning stage you need to consider only the risks of pedestrians who may be struck by vehicles operating on the site

B Yes. You need to plan to avoid disruptions to the local traffic in the area and make sure you have the right materials at the right time

C No. You need make sure there are control measures to avoid crossing traffic, and segregate pedestrians and vehicles wherever possible

D Yes. During planning you need to consider the proximity of schools, hospitals and built up areas to the site, and possible busy periods

14.50 To minimise the risk of vehicle-related incidents during deliveries to site, which of the following is an effective practical control measure for the safe management of deliveries?

A Divert pedestrians away from the site and provide information for routes to follow

B Design storage compounds with enough space for delivery vehicles to turn round

C Locate storage and loading areas near the entrance and off primary vehicle routes

D Locate storage and loading areas away from areas of frequent pedestrian activity

14.51 Which one of the following should be considered at a project's planning stage to reduce the risk to pedestrians and drivers from site traffic entering or exiting?

A Provide trained marshals or signallers to control site vehicle movements

B Control material stocks to ensure you have the right levels at the right time

C Implement diversions for pedestrians away from site vehicle traffic routes

D Regular meetings to inform the community of site traffic arrangements

14.52 You are planning a safe system of work for traffic management and vehicle movements on a new project. Which one of the following will help to effectively manage the risks occurring from delivery vehicles entering and leaving the site?

A Ensure delivery companies and drivers have the correct haulage licenses

B Ensure delivery companies have advance notice of unloading arrangements

C Ensure delivery drivers do not leave the cab of their vehicle while on site

D Ensure delivery drivers leave the site only when the vehicle is sheeted

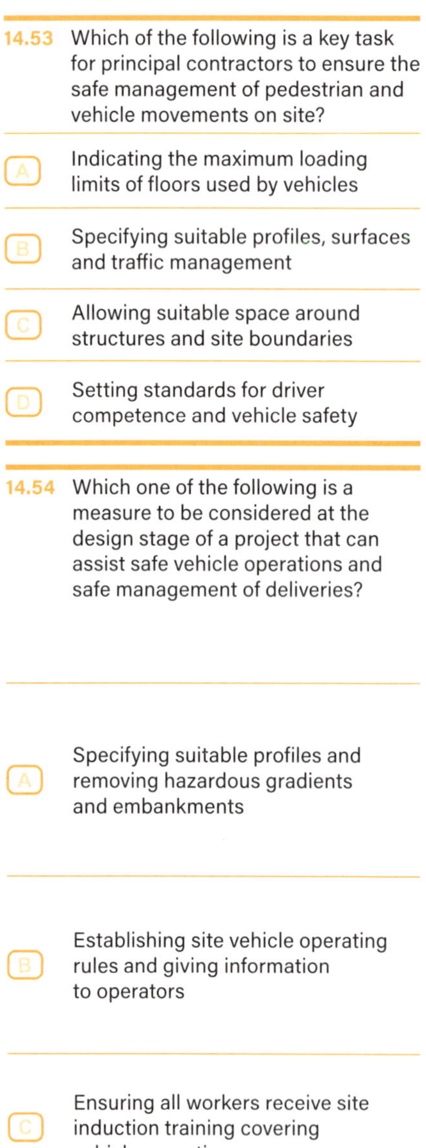

14.53 Which of the following is a key task for principal contractors to ensure the safe management of pedestrian and vehicle movements on site?

A. Indicating the maximum loading limits of floors used by vehicles

B. Specifying suitable profiles, surfaces and traffic management

C. Allowing suitable space around structures and site boundaries

D. Setting standards for driver competence and vehicle safety

14.54 Which one of the following is a measure to be considered at the design stage of a project that can assist safe vehicle operations and safe management of deliveries?

A. Specifying suitable profiles and removing hazardous gradients and embankments

B. Establishing site vehicle operating rules and giving information to operators

C. Ensuring all workers receive site induction training covering vehicle operations

D. Phasing works on site to minimise risks from vehicle operations to workers

14.55 When implementing a management plan to minimise the risk of vehicle-related accidents, which one of the following actions is most important?

A. Design compounds to allow delivery vehicles to turn round

B. Direct all traffic routes leaving site through the wheel wash

C. Discourage private vehicles from being parked near the site

D. Divert site vehicles and traffic away from built up areas

High risk activities

15 Working at height

15.01 Where a work at height risk assessment has identified a fall-arrest system as the last resort, what else must be in place before work can start?

A. Emergency rescue plan

B. Replacement harness

C. Double anchor point

D. Restraint lanyard

15.02 When compiling a rescue plan for operatives who may fall from height into a safety netting system, which **three** measures must be considered?

A. Provision of full first aid with trained personnel

B. Access at the bottom of the netting

C. Provision for emergency recovery for operatives

D. Recording the details of the individual weight of the operatives

E. Access from height to aid quick recovery

15.03 According to the Health and Safety Executive, what is an essential part of fall protection systems?

A. Method statement

B. Rescue plan

C. Protection zone

D. Certified workers

15.04 A roofer has fallen from height leaving them suspended in a conscious and uninjured state. Which two of the following statements are true regarding their rescue?

1	There is ample time to plan and carry out the rescue as they are safe in their harness	A	1 and 2 only
2	They should be rescued promptly to prevent fainting and unconsciousness	B	2 and 3 only
3	They should refrain from moving and remain still while they await rescue	C	3 and 4 only
4	They should gently exercise and move their legs to slow suspension syncope	D	2 and 4 only

15.05 Why is it important to purchase fall protection equipment along with a specialist kit?

A To comply with the Work at Height Regulations 2005

B To allow adjustment to the individual's body weight

C To aid the rescue of individuals suspended at height

D To make the rescue easier for the emergency services

15.06 Which one of the following can cause the onset of 'suspension syncope'?

A Working on a steep pitched roof with an anchored lanyard for five hours

B Falling from a height and hanging in a harness for more than ten minutes

C Carrying out a manual task for thirty minutes from an aluminium ladder

D Falling from a height using a proprietary suspension loop and hanging for 10 minutes

15.07 When rescuing an injured person who has fallen into a safety net, is it true that no other injuries can be sustained during rescue?

A Yes, the victim is safe in the net because it is supporting them and keeping them secure while awaiting rescue

B No, further injury to the victim could be unavoidable due to the stretching nature of the net as a rescuer enters

C Yes, the tight knit structure of the net removes the chance of further injury to the victim from other potential hazards

D No, the victim cannot be rescued from the net, so they have to attempt to make their own way out

15.08 What is the recommended maximum height for a free-standing mobile tower when used indoors?

A. As specified by the person erecting the access equipment

B. As specified by the site health and safety advisor

C. As specified by the access equipment manufacturer

D. As specified by the working at height regulations

15.09 What is the main significance of using a proprietary suspension loop along with a fall arrest lanyard when working at height?

A. It would allow a conscious casualty to rescue and lower themselves to the ground

B. It would prevent the conscious casualty becoming entangled within their own harness

C. It would allow a conscious casualty an easier rescue using this additional line

D. It would lessen the effects of suspension syncope on a conscious casualty

15.10 A worker has fallen and been arrested by their fall protection equipment. What is the main reason for a quick rescue response?

A. To prevent the onset of fainting

B. To stop them from having a cardiac arrest

C. To prevent them from being concussed

D. To avoid added weight on the anchor point

15.11 Which **two** of the following factors must be considered when selecting the correct type of mobile tower?

A. That it is suitable for the type of work to be carried out

B. That there is a suitably large enough area to erect it in

C. That it will have a suitably embedded anti-static device

D. That it is suitable for the environment that it will be used in

E. That it is constructed from suitable lightweight aluminium

15.12 A painter is working from the platform of a mobile access tower. To prevent the operative from disembarking the tower to replenish paint materials, they are being hoisted up the outside of the tower. Which one of the following could be an outcome if not controlled sufficiently?

A It could create limited space on the working platform

B It could cause the tower to sway from side to side

C It could significantly damage the tower assembly components

D If could impact stability and cause the tower to overturn

15.13 Electricians are using a tower scaffold to repair ceiling lights in a factory and it has been necessary to alter its position and remove the guard rails to get around obstacles. What must happen before it is used again?

A The parts are placed back in position as soon as possible by the person who removed them

B Pre-use checks are completed and recorded by a competent person on the inspection tag

C All the parts removed are checked for any damage before replacing them back again

D The structure is dissembled then rebuilt by a competent person to the manufacturer's instructions

15.14 Before using a scissor lift on site, an audit of the operator's competency card confirms that they hold a valid PASMA qualification. Is this the correct qualification to operate this type of equipment?

A Yes, this is the correct qualification to hold from the Prefabricated Access Suppliers' and Manufacturers' Association (PASMA)

B No, this is the wrong qualification as they should hold an International Powered Access Federation license (IPAF)

C Yes, this is an accepted qualification but they must also hold an International Powered Access Federation license (IPAF) 3b + card

D No, this is not the correct qualification the user would need to attend a training course to gain a Push Around Verticals' (PAV) license

15.15 An operative has been asked to operate a boom platform but they have not operated one recently. They do hold a valid International Powered Access Federation license (IPAF), but what else must they undertake before operating the equipment?

A A refresher IPAF course so their license is renewed due to the lapse of time since operating this equipment

B A full health assessment to deem satisfactory fitness levels with a qualified health practitioner

C A walk-round familiarisation check of the equipment with a competent engineer to refresh knowledge of the model

D A familiarisation training session on the model over a realistic time period with a competent person

15.16 An operative has been asked to use a scissor lift to carry out some urgent work at height but they are not competent to use this type of equipment. Which of the following must the operative do before they can operate it?

1	Receive familiarisation training sessions from a competent person	A	1 and 2 only
2	Download and use a computer simulator app for scissor lifts	B	2 and 3 only
3	Receive formal training by a recognised training body	C	3 and 4 only
4	Hold a correct valid training card or certificate	D	1 and 4 only

15.17 An engineer qualified to operate a mobile elevated work platform is working in a high-risk area of entrapment, near overhead steel beams. What must be done to reduce this significant risk to the worker before they start this work?

A Check that the equipment has a primary guard rail installed

B Select equipment that has a secondary guarding device installed

C Install height restrictor devices on the platform controls

D Make sure that the operative is issued with protective body armour

15.18 Which one of the following must employers consider first in regard to work at height activities?

A The consequences of falling

B All equipment is inspected

C Whether it can be avoided

D Use of appropriate equipment

15.19 To comply with the Work at Height Regulations 2005, employers must ensure the safety of those working at height, and who else?

A Visitors who are being supervised

B Plant operators working nearby

C Individuals who are passing below

D Contractors working around the site

15.20 Extensive roof repairs are being carried out on a gable end of a domestic property. The contractor has deemed a ladder as the most suitable way to access the roof. Is this complying with the Work at Height Regulations 2005?

A Yes, because the contractor has risk assessed the work and deemed this the most suitable access equipment.

B No, because all work from ladders has been banned and the contractor needs to provide a scaffold for this work.

C Yes, because the ladders can still be used to carry out some work as long as the ladder has been inspected.

D No, because the contractor must reduce the risks further and provide a suitable and sufficient safe means of access.

15.21 Which of the following two work situations are covered by the Work at Height Regulations 2005?

1	Standing on the back of a lorry to unload materials	A	1 and 2 only
2	Carrying out a task in an excavation or shaft	B	2 and 3 only
3	Making an additional entrance in an internal wall	C	3 and 4 only
4	Operating a forward tipping dumper	D	1 and 4 only

15.22 In line with the Work at Height Regulations 2005, which one of the following is the most effective way to avoid working at height?

A. Contractors manage their own risks for working at height and use suitable access equipment

B. Architects and designers design out the need to work at height in the design phase

C. Employers refuse any work that requires extensive work at height procedures

D. Clients contract out the work to professional work at height operatives

15.23 Which **three** of the following requirements of the Work at Height Regulations 2005 must be considered when planning and undertaking work at height?

A. The weather conditions that could compromise worker safety

B. The cost of access equipment required for the work

C. The storing of materials safely so they won't cause injury if disturbed

D. The wearing of correctly fitted harnesses and lanyards before work starts

E. The arrangements for dealing with emergencies and rescue

15.24 What is the main purpose of the Work at Height Regulations 2005?

A. To prevent death and injury from any falls from height

B. To provide enforcing rules and regulation for any work at height

C. To preserve the health and safety standards for all work at height

D. To promote a culture of authority and standards for all work at height

15.25 What is the main difference between an employer's obligations and those of their employees when meeting the requirements of the Work at Height Regulations 2005?

A Employees must assess and record the risks of any work-at-height process, and inform their employer

B Employees must co-operate with their employer to make safe suggestions for work-at-height access

C Employers must assess the risks of working at height, and employees must co-operate with them

D Employers must inform employees of a structured approach, and co-operate when working at height

15.26 When carrying out a risk assessment on a work-at-height activity, why is it important to understand the concept of short duration work?

A To determine how many people will be needed to complete the task

B To determine the amount of personal protective equipment required

C To determine if ladders are acceptable to complete the task

D To determine if fall-arrest provision will be needed to complete the task

15.27 To comply with the Work at Height Regulations 2005, several controls must be put in place. Identify **three** of these from the following.

A Where reasonably practicable, avoid the need to work at height

B Make sure appropriate access equipment is used to prevent falls

C Ensure all operatives wear personal fall protection equipment

D Minimise the distance and potential consequences of any fall

E Erect collective safety barriers to ensure protection of the public

F Make sure that ladders or stepladders are available for use

15.28 According to the hierarchy of risk control, what is the first consideration when assessing the risk of working from height?

A Avoid working from height where possible

B Select the correct access equipment

C Make sure workers are appropriately trained

D Prevent falls from occurring

15.29 What is the first hierarchy of risk control when working at height?

A Avoid the need to work at height

B Prevent the risk from falling

C Provide appropriate access equipment

D Install collective fall protection

15.30 A newly qualified roofer has been asked to carry out slate repairs to a domestic property using a roofing ladder. What should be their first consideration?

A Allocate this work to someone else with more experience

B Ensure the ladder complies to the British Standard EN131

C Change the access equipment to a collective protection type

D Ensure that the ladder can be properly secured top and bottom

15.31 When a work at height task cannot be avoided, which one of the following is the next consideration?

A Select the correct access equipment

B Provide appropriate personal protection

C Minimise the collective risk from falling

D Ensure workers are trained and competent

15.32 When complying with the Work at Height Regulations 2005, the hierarchy of risk control puts personal protection equipment above collective protection measures. Is this correct?

A Yes, collective protection should be considered after individual risks have been identified

B No, there is no specific level of precautions, this depends on the findings of the risk assessment

C No, personal protection should be considered after collective risks have been taken into account

D Yes, personal protection systems that take into account everyone at risk are always the first option

15.33 A joinery team is erecting a timber frame roof and cannot avoid the risk of working from height. Which of the two procedures must the team follow to control any further risks from this work?

1	Provide all workers with a current risk assessment they must follow	A	1 and 2 only
2	Erect a suitable scaffold with edge protection around the working area	B	2 and 3 only
3	Install collective safety netting under the timber frame roof area	C	3 and 4 only
4	Fit suitable barriers around ground level to stop unauthorised access	D	1 and 4 only

15.34 A worker is using a mobile elevated working platform to install solar panels on an external pitched roof. Applying the hierarchy of risk control, what must be put in place next for this type of work?

A Completion of a plan so workers can be rescued in the event of an emergency

B Fitting of toe boards to the access equipment to stop tools being accidentally kicked over

C Issuing of short-work restraint lanyard secured to the manufactured anchor point within the basket

D Installation of collective safety nets or air bag systems close to the level of work

15.35 A joiner is installing new roof windows into a second-floor domestic property. How can they significantly reduce the risk from working at height?

A Install a scaffold around the roof with additional edge protection

B Carry out the work from a mobile operated platform with extra barriers

C Use lifting equipment that can take the materials up to the working height

D Carry out the task from within the property and below the roof

15.36 When working at height in line with guidance LA455 Safe Use of Ladders and Stepladders, which one of these criteria must be followed?

A Work must be completed within a time frame of 2 hours

B The worker must have two hands free at anytime to carry out the task

C Work being carried out must not compromise the stability of the ladder

D The ladder needs to be positioned so the user is side-on to the work

15.37 Which one of the following is not a safe way for using a ladder?

A Resting it on a gutter with one rung clearance

B Positioning it on a firm and level surface

C Inspecting it before use to check for defects

D Positioning it away from pedestrian routes

15.38 British Standards and regulations for ladders changed in 2017 to BS EN 131. This means that any ladders that do not meet this new standard are deemed unsafe and should be destroyed. Is this correct?

A Yes, if a company is found to be using any ladders meeting only the previous standard, they will be prosecuted by the Health and Safety Executive

B No, ladders can still be used meeting the previous standard as long as they are in good condition and inspected regularly

C Yes, because previous standard ladders were all destroyed by the end of 2017 and new compliant ladders purchased

D No, a company should not use any ladders under the previous standard because they are too narrow at the base and were deemed unsafe

15.39 Which one of the following do you need to be aware of when using telescopic ladders?

A They are difficult and awkward to extend

B They can easily blow over in the wind

C They are prone to twisting and flexing

D They become very rigid when fully extended

15.40 What is the first consideration before erecting a ladder?

A The wind level is within satisfactory parameters

B There is another person available to assist

C The ladder is long enough for safe access

D There are no overhead obstructions

15.41 An operative requires a ladder for a short duration task. A full risk assessment has been conducted and identified that a two-piece ladder is the safest method of access. Before the ladder can be used, which two of the following must the operative do?

1	Check an up-to-date and sign-off safety tag is attached to the ladder	A	1 and 2 only
2	Measure the ladder to ensure it is long enough for the height to be reached	B	2 and 3 only
3	Undertake a visual pre-use check of the ladder	C	1 and 3 only
4	Check the ladder is colour coded to identify its weight limit	D	2 and 4 only

15.42 Which one of the following is an example where it is acceptable to 'foot' a ladder?

A When it is being used for short term work and constantly being moved as the work progresses

B When it is being climbed for the first time for the purpose of tying and securing it at the top

C When it is not possible to get a secure fixing at the top due to the nature of the material it is lying on

D When it is being used on its own for one worker and not part of an access scaffold

15.43 When completing a ladder inspection you notice a dent in one of the rungs. Which one of the following must you do?

A Take the ladder out of service and report it

B Update the ladder tag to record the defect

C Repair the ladder immediately

D Use the ladder as the task is urgent

15.44 When a mobile elevating work platform is being used for lifting people, the frequency for thorough examination should not exceed...

A 3 months

B 6 months

C 9 months

D 12 months

HIGH RISK ACTIVITIES

15.45 What operations of a mobile elevating work platform must the operator be familiar with before its use?

A The seatbelt adjustments

B Turning it 80 degrees

C Elevation to the maximum height

D The emergency system

15.46 When selecting a mobile elevating work platform, which one of the following would not be a priority for consideration?

A The type of ground for travel and positioning it

B The use of a marshal to assist with manoeuvring

C The type of beacon fitted for site-traffic management

D The location of overhead power lines

15.48 When operating a mobile elevated work platform at height, when is it acceptable to lower the platform using the ground controls?

A Only when the person in the basket has instructed an individual on the ground

B Only if the operator in the basket is unable to, due to an emergency situation

C Only if the operator in the basket is having difficulties going around an obstruction

D Only if the person in the basket does not have full vision of the working area

15.47 An engineer is carrying out an inspection from a mobile elevated work platform over a flowing river. Is it acceptable for them not to be clipped to it?

A No, because it is stipulated in the safe working procedures that individuals must be anchored to the machine by a harness

B Yes, because the risk assessment identified that the risk of drowning was greater than the fall, but a life vest must be worn

C No, because there is still a risk of falling into the river, and when the fall is arrested a rescue will be much faster

D Yes, because the risk assessment identified that the boom cage provided sufficient protection if a fall occurred

15.49 Why must the safe working load of a mobile elevated working platform not be exceeded when being used?

A Because It would not be able to reach full working height

B Because this would cause the tyres to deflate

C Because it would create instability when fully extended

D Because it would be unable to gain traction over uneven ground

15.50 When carrying out end of day security inspections, it is found that workers have left the keys in the mobile elevated working platform. What would be the main risk of this?

A The machinery brakes will be able to unlock and it could roll away causing damage

B The digital log of use on the machinery will keep running, extending the leased hourly rate

C The fuel to the machinery will remain on trickle and could pose a fuel leak and fire risk

D The machinery could be accessed by unauthorised persons causing vandalism

15.51 When selecting a fall protection system to allow work to be conducted safely from a boom-type mobile elevated work platform, which two would be correct?

1	Use of a work-restraint system as opposed to a fall-restraint system	A	1 and 2 only
2	Attachment of a lanyard to a designated anchor point inside the work platform	B	3 and 4 only
3	Use of a fall-restraint system as opposed to a work-restraint system	C	2 and 3 only
4	Attachment of a lanyard to a designated anchor point outside the work platform	D	1 and 4 only

15.52 When operating a mobile elevated working platform (MEWP) in a restricted or enclosed space, which **three** of the following procedures must be adhered to?

A Operators have an in-depth understanding of all the controls

B The MEWP is fitted with a bumper system to avoid damage to walls

C Battery operated MEWPs are not recharged within the space

D Liquefied petroleum gas powered MEWPs are re-fuelled outside the space

E Operators have all been extensively trained in emergency first aid response

15.53 A mobile elevated working platform (MEWP) is being operated along a building perimeter that runs adjacent to a busy high street. Segregation is in place between pedestrians and the machine. Which one of the following precautions must also be in place?

A Fluorescent strips are attached to the machine to ensure that it is visible

B The machine positioning does not compromise the view of street signs

C Noise levels of the machine are restricted to reduce the impact of noise pollution

D Additional guards are fitted to the machine to prevent unauthorised tampering

15.54 You have been asked to carry out roof repairs on a surface containing cement sheeting. What is the most significant risk?

A You could slip on the surface area of the sheeting due to a build up of moss

B You could fall through the sheeting, as it is a highly fragile surface

C The sheeting could shatter as it is brittle, damaging the roof surface

D Touching this type of surface with bare skin could cause dermatitis

15.55 When work on a fragile surface cannot be avoided, which one of the following must be implemented to mitigate the risk from falls?

A Verbally inform everyone within the working area of the hazards from the fragile roof

B Ensure suitable and sufficient platforms, guard-rails and load-bearing covers are installed

C Cover the dangerous fragile area with safety netting and tell everyone to be careful

D Mark off the dangerous fragile area with red and white warning tape and install signage

15.56 When assessing if a roof is safe to work on, which one of the following applies?

A If the roof's pitch is less than 30 degrees and covered in slate, it is safe to walk on

B All roofs should be treated as fragile until a competent person has confirmed otherwise

C A roof should be visually inspected by a roofer from ground level and confirmed it is safe

D All roofs should be accessed safely using the correct roofing ladders supported on ridge tiles

15.57 Repairs need to be carried out on fragile roof lights. What is the safest way to access the roof lights?

A From the top using a mobile elevated work platform

B From the side using a suitable roofing ladder

C From underneath using a suitable working platform

D From the side using roof boarding and guard rails

15.58 By fitting which one of the following to a roof light area will demonstrate that a building owner has considered the hazards of working on fragile roofs?

A Warning tape around the edge of the roof light

B A metal frame cover fitted over the roof light

C Edge protection fitted to the edge of the roof light

D Collective netting installed under the roof light

15.59 A worker has been asked to install solar panels on a roof. Which of the following **three** surfaces should be treated as fragile, requiring extra safety precautions to be taken?

- A Concrete
- B Slates and tiles
- C Corroded metal sheets
- D Wired glass
- E Steel
- F Timber

15.60 A plumber needs to work on a slate extension roof to install an extractor duct. Will the roof be strong enough for this work?

- A No, this type of material is unpredictable and can collapse when added weight is put on it
- B Yes, this type of material is strong enough for a person to stand on but for short periods of time only
- C No, this type of material should not have been used on a roof and is not suitable for this work
- D Yes, this type of material is very strong, robust, and able to withstand this work

15.61 What significant duties must a roofer complete while carrying out lighting repairs on an industrial roof?

1	Sign off the repairs safe system of work and return this to the supervisor	A	1 and 2 only
2	Leave a mark on the external wall indicating the repair made above	B	2 and 3 only
3	Mark the fragile parts of the repair using red screws or bolts	C	3 and 4 only
4	Record details of the repairs in the health and safety file for the building	D	1 and 4 only

15.62 When planning a safe system of work for two similar pitched roofs (one is slate and tile, the other is fibre-cement sheets), what should be the main safety consideration?

A The workers can only walk on the bolts of the fragile sheet roof, but can stand anywhere on the tiled roof

B The workers will need crawling boards installed on the fragile sheet roof, but can walk directly on the tiled roof

C The workers should work from a roofing ladder on both roofs, so long as it has the correct safe working load

D The workers will need suitable protection methods installed on both fragile roofs, to prevent anyone standing on them

15.63 What is the main reason for using a safety net system rather than a personal fall-arrest system?

A They are cheaper than personal equipment

B They allow easier rescue in the event of a fall

C They can cover a wide floor area

D They provide collective protection from falls

15.64 A safety netting system eliminates the risk of a worker falling from height and injuring themselves? Is this true?

A Yes, the rigid structure of the netting will arrest the fall and prevent any injury

B No, the netting is designed only to catch fallen materials and debris

C Yes, the netting is designed to reduce the likelihood of falling and to minimise any injury

D No, netting could reduce the severity of injury, but does not remove the likelihood of falling

15.65 How would the user of a safety netting system ensure its suitability before use? Select **three** from the following.

A The label displays the maximum weight

B It has a unique serial number on the label

C It has been tested if older than 12 months

D The date of manufacture is visible on the label

E The mesh size is greater than 110mm

15.66 To comply with the Work at Height Regulations 2005, safety netting should allow a fall distance from the working platform to be.....

A as little as reasonably practicable

B enough to allow a person to stand up

C 6 m below the working platform

D twice the height of the average worker

HIGH RISK ACTIVITIES

15.67 After a safety netting system has been installed which one of the following procedures must be completed before use?

A It is checked and inspected by a competent person

B It is tested by workers via a simulated training session

C It is double anchored at every attachment point

D It is secured underneath with a finer net to collect debris

15.68 How often should safety nets older than 1 year be tested?

A Within the last 18 months

B Within the last 12 months

C Within the last 6 months

D Within the last 3 months

15.69 What is the significant difference between a 'knotted' and a 'knotless' safety netting system?

A A knotted type has more elasticity and bounce to take a fall compared to that of the knotless type

B A knotless type needs regular maintenance checks whereas the knotted type does not

C A knotted type is heavier and older whereas the knotless type is newer and made from plastic

D A knotless type must be securely anchored at four points whereas the knotted type does not

15.70 When installing safety netting for workers protection, which two of the following should be applied?

1 Ensure it is erected directly in line with the end of the working platform

2 Ensure it is erected as close as possible to the working level

3 Ensure the installation area is free from obstructions

4 Ensure it is registered and logged in the plant maintenance records

A 1 and 2 only

B 2 and 3 only

C 3 and 4 only

D 1 and 4 only

15.71 Why is it important to make sure that safety nets have sufficient clearance below them?

A To allow access to the net from below for rescue equipment in the event of an emergency

B To stop the net from making contact with the ground as it will deform after installation

C To allow adequate clearance for the safe passage below the net of plant and equipment

D To stop a fallen person from striking the ground due to the deformation that will occur in the net

15.72 What is the best way to check that contracted scaffolders are competent to carry out work on site?

A Ask to see their identification card showing their qualifications

B Ask their manager for a list of previous jobs completed

C Ask them some questions to check their level of knowledge

D Ask to see the risk assessment and method statement for the work

15.73 When selecting a scaffolder to complete work on site it is important to ensure their competence. This can be done by checking which **three** of the following?

A Their public liability insurance certificate

B Their membership of a recognised trade association

C Their health and safety performance records

D Their tools and equipment quality

E Their evidence of completed previous work

15.74 The Work at Height Regulations 2005 require scaffolders to demonstrate which one of the following?

A That they have the required knowledge to select the correct equipment needed

B That they have the required training to erect, dismantle and alter the equipment

C That they understand well how to follow the manufacturers guidelines

D That they have had sufficient training to complete the risk assessment adequately

15.75 Where would a competent scaffolder find information to ensure their scaffolding design is compliant?

A In the TG20:21 operational guide

B In the Work at Height Regulations (2005)

C In the BS EN 131 standard

D In the NASC SG4 guide

15.76 Roofers have taken delivery of a large quantity of roof tiles. Why must the scaffolders be informed about these additional materials?

A To ensure the working platform is not overloaded

B To allow for alterations of the platform width

C To temporarily remove the guard rails for the material loading

D To install additional hoists for safe loading of the materials

15.77 What is the main significance of having a bespoke scaffold design erected on site?

A It would have a greater cost due to the extra materials and added checks required

B It would require specialist trained operatives to install and maintain it throughout

C It would need a higher level of health and safety checks and recording systems in place

D It would require strength and stability calculations completed before being erected

15.78 When scaffolders erect a tube and fitting structure complying to TG20:21 and BS EN 12811, additional calculations do not need to be carried out. Is this correct?

A No, all standard solutions still need additional calculations considering the configurations

B Yes, as they provide worked out calculations for all scaffolding types, including system structures

C No, they still require additional calculations to be carried out on a tube and fitting structure

D Yes, as they provide recognised standard configurations with approved calculations

15.79 A scaffolding structure has been modified. Which two of the following must a competent person carry out following this modification?

1	An inspection and report	A	1 and 2 only
2	An equipment check and audit	B	2 and 3 only
3	Removal of all inspection tags	C	3 and 4 only
4	A handover certificate	D	1 and 4 only

15.80 A severe storm has passed over the site. What must now happen in relation to safe working on the scaffolding structure?

- A Workers need to be warned that the platform will be slippery
- B Inspection tags need to be checked to ensure they are still attached
- C The structure needs to be checked by a competent person for deterioration
- D The safe system of work needs to be updated and re-issued

15.81 Which of the following statements is true regarding roofing ladders?

- A They should extend at least 2 metres above the eves
- B They should only be used for short duration or low-risk work
- C They should be tied together with the access ladder
- D They should be positioned with the anchor hook pushing on the ridge tile

15.82 Which one of the following statements is false in relation to the safe use of stepladders?

- A The guard-rail barrier should be closed when in use
- B The retaining hinges should be equal in length
- C The legs should be positioned as far apart as the mechanism allows
- D The user should be able to work from the top three steps

15.83 What is the main difference between podium steps and stepladders?

- A Podium steps have a restriction on the working height
- B Stepladders allow the worker to position themselves closer to the work
- C Podium steps allow the worker to face any side of the working platform
- D Stepladders are much easier to store and inspect

15.84 When setting up lightweight staging for access to work, which one of the following must it not be?

A Placed on a firm and level base

B Fitted with toe boards and guard rails

C Marked with the maximum permitted load

D Accessible using a second platform

15.85 Identify the **three** significant failings associated with incidents involving tower scaffolds.

A Ignoring the system of work

B Referring to the manufacturer's information

C Moving the frame with persons on the platform

D Using parts from other systems

E Using the through-the-trap motion (3T)

15.86 A mast climbing work platform (MCWP) needs to be used to fit external cladding to a high-rise building. Which two significant procedures must be in place before use?

1 Make sure that you can identify a loss of mechanical integrity in each drive unit

2 Check each drive unit is fitted with a mechanical device such a centrifugal brake

3 Make sure each drive unit contacts the buffers and braces with every descent

4 Check all switches such as limit ones are positioned at the highest acceptable level

A 1 and 2 only

B 2 and 3 only

C 3 and 4 only

D 1 and 4 only

15.87 A mobile elevating working platform (MEWP) is required for workers to erect an external sign to a factory building. There is a live electrical cable within a 5 metre radius of the work location, but there is no risk to the operators from electrical shock. Is this true?

A Yes, the safe working distance from a live electrical cable is under 4 metres, so there is no electrocution risk in this work

B No, the risks from electrocution need to be managed and controlled as the live electrical cable is within a distance of 10 metres

C Yes, all live electrical cables are fitted with insulation and earthing devices to stop arcing of electricity so there is no foreseeable risk

D No, there is a risk of electrocution to workers but only if they make contact with the live electrical cable by touching it with their tools

15.88 What significant difference exists between using mast climbing work platforms (MCWPs) opposed to mobile elevating work platforms (MEWPs) that can influence its selection for use on a work at height project?

A They allow several operatives to work off a single platform at the same time at the desired height

B They need two or more drive units to co-ordinate together to raise the platform to the required height

C They provide protection to the risks of falling while providing temporary access to work at height

D They require trained operatives to hold the required certificated level to use them for work at height

15.89 What is the main significant operating difference between a boom-type and scissor-type mobile elevating working platform (MEWP) that needs to be considered before selecting one for a specific activity?

A The scissor lift mechanisms allow for a higher reach with heavy bulky materials

B The booms' structure is less suitable to use for installing long heavy materials

C The scissor-type is more suitable for rough and uneven ground surface terrain

D The boom-type can be safely and effectively operated by one person

15.90 Which one of the following checks would not be carried out before using stilts to apply plaster to ceilings in a domestic house?

A Equipment is in good working condition

B Workers have received training on their correct use

C Apparatus meets the current British Standard

D Floor surface is level and free from obstructions

16 Excavations and confined spaces

16.01 You are delivering a safety brief to operatives who are about to conduct work in a sewer chamber. Which one of the following hazards must you talk about?

A The possible presence of hydrogen dioxide, which even if inhaled in small quantities can cause serious burns and death

B The possible presence of hydrogen sulphide, which if inhaled can cause convulsions and cardiac arrhythmias

C The possible presence of hydrogen monoxide, which if inhaled in small quantities can cause internal tissue damage and death

D The possible presence of hydrogen peroxide, which if inhaled can cause permanent lung damage or pulmonary edema

16.02 The use of any petrol- or diesel-powered equipment within a confined space creates an excess of carbon dioxide. Is this correct?

A Yes. Petrol and diesel engines create an excess of carbon dioxide, a suffocating hazard which can result in cardiac arrhythmias and death within minutes

B No. Petrol and diesel engines create concentrations of hydrogen sulphide which can result in collapse, respiratory paralysis and death in a short time

C Yes. Petrol and diesel engines create concentrations of hydrogen dioxide which is odourless and tasteless, resulting in collapse and death within minutes

D No. Petrol and diesel engines create carbon monoxide, which is an extremely toxic gas causing loss of consciousness and death in a very short time

16.03 Which one of the following is the main risk from an oxygen-enriched atmosphere within a confined space?

A Ordinary materials such as paper and clothing will burn with exceptional ferocity

B Extremes of heat will lead to a dangerous increase in body temperature

C Can cause loss of consciousness and death in a very short time

D A toxic or suffocating hazard and a hostile environment may exist

16.04 When carrying out a risk assessment for working in a confined space, which one of the following needs to be considered as a potential hostile environment danger?

A Use of any form of internal combustion engine

B Leakage from valves, flanges or blanks

C Vapours from solvents, white spirit or thinners

D Oxy-propane welding and other hot works

16.05 Which one of the following oxygen levels is the normal acceptable minimum level for working in a confined space?

A 21%

B 19%

C 17%

D 15%

16.06 Which one of the following is an example of major sources of explosive and flammable hazards in confined spaces?

A Hydrogen

B Nitrogen

C Ammonium

D Sodium

16.08 Acute exposure to high concentrations of hydrogen sulphide can result in which **two** of the following?

A Light-headedness and increased breathing difficulties

B Nausea, respiratory difficulties and possible collapse

C Cardiac arrhythmias and death within minutes

D Respiratory difficulties and loss of consciousness

E Respiratory paralysis, cyanosis and convulsions

16.07 Which one of the following is the correct action to take first if excess oxygen is discovered in a confined space?

A The space must be quickly ventilated and the source of the oxygen leak identified

B The workers should ensure there are no sources of ignition present within the space

C The workers must not enter the space unless they are wearing breathing apparatus

D The space must be quickly evacuated and ventilated until normal levels of oxygen are regained

16.09 Plans and drawings provided by utility companies are an accurate way of identifying the location of underground services. Is this correct?

A Yes - utility companies use ground-penetrating radar equipment to identify accurate locations of services on plans and drawings

B No - changes in the colour of the surface material, such as recently-laid tarmac indicate where a service trench can be found

C Yes - utility companies' plans and drawings have co-ordinates, and lines and levels of newly-installed services are included

D No - trial holes should be dug using hand excavation methods to establish the exact location and depth of the services

16.10 Plans and drawings showing the locations of underground services should be considered as historical documents that cannot be fully relied upon, and you should be careful when reading them. Is this correct?

A No - utility companies' ground-penetrating radar surveys are used to identify accurate locations of services and connections on plans and drawings

B Yes - reference points may have been moved, surfaces regraded or services moved without authority, and not all connections may be shown

C No - once routes have been identified, the co-ordinates, lines and levels of newly-installed services and connections are included

D Yes - changes in colour of the surface material, sometimes known as tracks, will indicate where a service trench can be found

16.11 The Electricity at Work Regulations (1989) require that live overhead power lines above 33,000 volts must be isolated and made dead to prevent danger before any work takes place. Is this correct?

A Yes. Power lines should be isolated and made dead and practical steps should be taken to prevent danger from any live cable or apparatus

B No. Where work cannot be avoided the local electrical company should be consulted before any work is started, and a safe system of work planned and implemented

C Yes. Power lines should be isolated and made dead or where necessary, wait for the supplier to re-route them to enable the work to take place

D No. Work may only be carried out in close proximity to live overhead lines when there is no alternative, and only when the risks are acceptable and can be properly controlled

HIGH RISK ACTIVITIES

16.12 Which one of the following is the most common injury resulting from strikes to underground services?

A Damage to gas pipes, resulting in burns from fire or explosion

B Contact with electrical cables, resulting in major burns or death

C Damage to fibre optic cables, resulting in burns to the eyeball

D Contact with raw sewage, resulting in skin burns or leptospirosis

16.13 A conventional cable avoidance tool (CAT) will not locate which one of the following?

A Electricity cables

B Telecommunications cables

C Lead water services

D Plastic gas pipes

16.14 When using a cable avoidance tool (CAT) to conduct a survey for the presence of underground services, the radio frequency is mainly used to detect which one of the following?

A Fibre optic cables

B Telecommunications cables

C Gas and water mains

D Drains and sewer pipes

16.15 When using a cable avoidance tool (CAT) to locate underground services, a generator can be attached to an exposed part of a pipe or cable for which one of the following reasons?

A To detect hidden flat metal covers, joint boxes, and so on

B To sweep a wider area where there are underground services

C To provide a signal for the CAT to track the line of services

D To provide a signal for the CAT to accurately locate plastic pipes

16.16 Which one of the following is a requirement of the Electricity at Work Regulations (1989), when work is required in an area where there are live overhead power cables?

A Where work cannot be avoided, consult the local electricity company before any work is started: a safe system of work must be planned and implemented

B Work may only be carried out in close proximity to live overhead lines when there is no alternative, and only when the risks are acceptable and can be properly controlled

C The live overhead lines should be isolated and made dead or, where necessary, wait for the supplier to re-route to enable the work to take place

D The live overhead lines should be isolated and made dead, and practical steps should be taken to prevent danger from any live cable or apparatus

16.17 At which **three** of the following intervals must an excavation be inspected?

A At the start of every work shift

B Following alterations to the safe system of work

C At three hourly intervals during a shift

D After any event likely to affect strength and stability

E After materials have fallen or been dislodged

F As part of the site's daily health and safety inspection

16.18 Which **two** of the following must be in place to enable work to be carried out safely in a confined space?

A Health and safety file

B Adequate rescue facilities

C Sufficient welfare facilities

D Permit to work system

E Respiratory protective equipment

16.19 Which **two** of the following factors could hinder effective communication when working in confined spaces and may need to be considered in the risk assessment?

A Level of noise outside of the confined space which may or may not be associated with the confined space work

B Distance between the point of entry of the confined space and the location of the first-aid facilities

C Training for workers on the methods of communication to be used when working in confined spaces

D Physical nature of the confined space or the presence of substances that could reduce visibility

E Face-fit tested masks or respirators limiting the methods of communication that can be chosen

16.20 As part of a permit-to-work, you are conducting air testing in a confined space. Which of the following toxic gases are colourless, odourless and tasteless, and can cause loss of consciousness and death in a very short time?

1	Carbon dioxide	A	1 and 4 only
2	Hydrogen monoxide	B	1 and 2 only
3	Carbon monoxide	C	2 and 3 only
4	Hydrogen dioxide	D	1 and 3 only

16.21 Access to and egress from many confined spaces is achieved by lowering or raising a person vertically through the entry or exit point. In these circumstances, which of the following apply?

1	The tripod and hoist must not be regarded as lifting equipment and the Provision and Use of Work Equipment Regulations (PUWER) would apply	A	1 and 4 only
2	Safety harnesses and rescue lines must be regarded as lifting accessories	B	3 and 4 only
3	The tripod hoist or other type of winch must be regarded as lifting equipment used for lifting persons and, as such, requires thorough examination on a six monthly basis	C	2 and 3 only
4	Safety harnesses and rescue lines must be regarded as personal protective equipment	D	1 and 2 only

16.22 Which one of the following is an essential feature of a permit to work?

A Identification of contractors engaged to carry out work

B Identification of those carrying out the hazardous work

C Identification of competent rescuers for a specific job

D Identification of who may authorise particular jobs

16.23 Which one of the following is the most accurate description of the definition of a confined space?

A A space which is enclosed or confined, and where serious specified injuries can occur from reasonably-foreseeable specified risks

B A space with hazardous conditions inside, or a change in the degree of enclosure or confinement

C A space in which reasonably-foreseeable specified risks are present due to its enclosed and hazardous nature

D A space in which there arises a reasonably-foreseeable specified risk from hazardous substances or conditions

16.24 Which of the following is the main duty of an employer under Regulation 4 of the Confined Spaces Regulations 1997?

A Testing the atmosphere or sampling the contents of confined spaces from outside, using appropriate long tools and probes

B Considering what measures can be taken to enable the work to be carried out without the need to enter the confined space

C Modifying the confined space itself to avoid the need for entry, or to enable the work to be undertaken from outside the space

D Using remote visual inspection (RVI) to carry out examinations, but only if this will provide the same results and safeguards as entry would

Environment

ENVIRONMENT

17 Environmental awareness and waste control

17.01 What should you do if you see construction materials stored under a tree?

A) Nothing, trees are not protected and materials can be stored anywhere

B) Make sure that the materials are removed if the tree shows signs of damage

C) Make sure that the materials are removed and tree protection fencing is installed

D) Make sure that the tree is watered regularly until the materials are removed

17.02 If it is suspected that bats are roosting on a site, which **two** of the following statements are true?

A) Bats breed during the winter and their roosts are protected only at this time

B) A bat survey should be carried out by a qualified specialist

C) Bats hibernate in the summer, and cannot be relocated during this time

D) A bat roost can be destroyed, damaged or relocated if it is unoccupied

E) A bat licence holder is legally allowed to capture, handle or relocate bats

17.03 Which one of the following is not a priority species?

A) House sparrow

B) Grey squirrel

C) Dormouse

D) Common toad

17.04 What legal duty does the Wildlife and Countryside Act 1981 place on a site manager when Japanese knotweed is discovered on a site?

A) To ensure it is moved to an isolated part of the site

B) To prevent it spreading to the wild by cordoning it off

C) To prevent it spreading to the wild by excavating it

D) To ensure it is undisturbed as it is protected by law

17.05 A construction project must protect and preserve the ecology of the surrounding area. Which of the following are the key reasons for identifying and managing wildlife early in the planning stage of a project?

1	To avoid investigation and prosecution under the Biodiversity Act	A	1 and 2 only
2	To avoid investigation and prosecution under environmental legislation	B	2 and 3 only
3	To avoid being targeted by eco-warriors or animal protection groups	C	2 and 4 only
4	To avoid costly delays to the programme and possible loss of reputation	D	3 and 4 only

17.06 To comply with water management and pollution control policy, which one of the following indicates some of the actions required prior to work starting on site?

A Identify all existing site drainage systems, mark them on site plans, and seal all drains to prevent accidental entry of mud and silt

B Identify all existing site drainage systems, mark them on site plans, and fit filter systems on all drains to prevent waste and pollution entering the drainage system

C Identify all existing site drainage systems, mark them on site plans, and cover/protect all drains to prevent accidental entry of mud and silt

D Identify all existing site drainage systems. mark them on site plans, and install anti-pollution barriers around every drain, at least 1m from the edge

17.07 Under the Treasure Act 1996 (for England, Wales and Northern Ireland), which one of the following would be classed as treasure?

A Objects that contain at least 40% of gold or silver, and are at least 200 years old

B Objects that contain at least 10% of gold or silver, and are at least 300 years old

C Objects that contain at least 20% of gold or silver, and are at least 200 years old

D Objects that contain at least 30% of gold or silver, and are at least 300 years old

17.08 You are managing the removal of asbestos waste from your site. Which one of the following statements is correct regarding its removal?

A It must have a controlled waste consignment note, be transported by a registered waste carrier and taken to a suitably authorised facility within 24 hours

B It must have a hazardous waste consignment note, be transported by a registered waste carrier and taken to a suitably authorised facility within 24 hours

C It must have a controlled waste transfer note, be transported by a registered waste carrier and taken to a suitably authorised facility within 48 hours

D It must have a hazardous waste transfer note, be transported by a registered waste carrier and taken to a suitably authorised facility within 48 hours

17.09 Which final element is missing from this four-point pollution incident response process? Stop – Contain – Notify – ?

A Review

B Clean up

C Take action

D Monitor

17.10 BREEAM is a third-party environmental tool that is used to certify the sustainability credentials of a building. What does BREEAM stand for?

A Building Research Establishment's Environmental Assessment Method

B Building Research Establishment's Environmental Assessment Management

C Building Research Establishment's Environmental Assessment Manual

D Building Research Establishment's Environmental Assessment Monitoring

17.11 Which one of the following would not be considered a benefit of an environmental management system?

A To improve cost and efficiency savings

B To reduce the use of brownfield sites

C To improve performance and customer relationships

D To contribute to wider Government targets

17.12 Which one of the following documents must be used for hazardous waste transfers?

A) A waste environmental permit

B) A waste transfer note

C) A waste management licence

D) A waste consignment note

17.13 Which one of the following controls should be in place to reduce the risk of pollution from a designated concrete and cement washout area?

A) It should be at least 5 metres from a watercourse

B) It should flow straight into a foul water drain

C) It should be at least 10 metres from a watercourse or drains

D) It should be directed so the flow soaks into alkaline-lined containers

17.15 You have been asked to complete a waste transfer note. Which **two** of the following sections must be completed?

A) Section A2 - How the waste is contained

B) Section D3 - European waste catalogue (EWC) codes

C) Section B2 - Name of the unitary authority or council

D) Section C2 - Contaminated land: applications in real environments codes

E) Section B3 - The process giving rise to the waste

17.14 You have hired an asbestos contractor to remove cement sheets from site. Would these be classified as dangerous for carriage, and so need to comply with the Carriage of Dangerous Goods and Use of Transportable Pressure Equipment Regulations 2009 (CDG) and the Control of Asbestos Regulations 2012 (CAR)?

A) Yes - this type of asbestos is classed as fibrous, which means that particles can escape during carriage, and so both sets of regulations would apply

B) No - the fibres are bound in cement and so cannot escape during carriage, therefore you would only need to comply with the CAR

C) Yes - this type of asbestos is classed as highly hazardous and must be double bagged in UN-certified bags, so both sets of regulations would apply

D) No - the fibres are bound in cement and cannot escape during carriage, therefore you would only need to comply with the CDG Regulations

17.16 Under the Environmental Protection Act 1990, it is an offence to operate a regulated waste facility without the relevant permit. Which of the following are the current maximum penalties?

1	Fined up to £50,000	A	1 and 2 only
2	Imprisoned for up to five years	B	1 and 3 only
3	Imprisoned for up to four years	C	2 and 4 only
4	Fined up to £30,000	D	3 and 4 only

17.17 Which **two** of the following are not associated with brownfield sites?

A It is unlikely that they contain contaminated land

B It is likely that they contain contaminated land

C They are exempt from the Environmental Protection Act 1990

D They have had some form of previous development

E They can pose risks to human health and the environment

17.18 When managing contaminated land, which two regulators should be involved in discussions regarding permits and licences?

1	The Health and Safety Executive	A	1 and 2 only
2	The relevant Local Authority	B	1 and 4 only
3	The Environment Agency	C	2 and 3 only
4	The Contaminated Land Agency	D	3 and 4 only

17.19 Before working on contaminated land, the risk level of a site indicates that a personal hygiene unit is needed. Which one of the following is the minimum requirement?

(A) The unit should be divided into at least three areas that includes a dirty area, a washing and toilet area, and a clean area. The unit must be cleaned and decontaminated on a daily basis.

(B) The unit should be divided into at least two areas that includes a dirty area, and a clean area. The unit must be cleaned and decontaminated after each use.

(C) The unit should be divided into at least four areas that includes a dirty area, a washing and toilet area, an incineration area, and a clean area. The unit must be cleaned and decontaminated on a daily basis.

(D) The unit should be divided into at least two areas that includes a dirty area, and a clean area. The unit must be cleaned and decontaminated on a daily basis.

17.20 Which one of the following is required when using mobile plant for the remediation of contaminated soils?

(A) Environmental certificate

(B) Environmental permit

(C) Remediation permit

(D) Remediation certificate

17.21 Gas membranes are becoming a more common part of the building process, and must be protected during construction. Which one of the following will not help to protect the membrane's integrity?

(A) Toolbox talks about safe procedures

(B) Restricted access to the working area

(C) Correct personal protective equipment (PPE)

(D) Warning notices near the working area

17.22 Before construction work begins, the local authority has identified a site as formally contaminated. What action must be taken?

(A) Fence the affected area off, restrict access, erect adequate warning notices that advise that the area is dangerous and what the contaminant hazard is

(B) Fence the affected area off, restrict access, erect adequate warning notices and post information leaflets to all local residents to inform them that the site is dangerous

(C) Fence the entire site off, restrict access, and erect adequate warning notices that advise all members of the public that the site is dangerous

(D) Fence the entire site off, restrict access, erect adequate warning notices and post information leaflets to all local residents to inform them that the site is dangerous

17.23 You have recently conducted a remediation process on a brownfield site that was formerly a lead works. Which one of the following would be your primary action to identify whether the process has been successful?

- A Monitoring, testing and sampling of watercourses within a 100m radius
- B Monitoring, testing and sampling of the land within the site boundary
- C Monitoring, testing and sampling of the air quality within a 200m radius
- D Monitoring, testing and sampling of the immediate land, watercourses and air

17.24 Should different classifications of soil be stockpiled separately?

- A No, all soils can be stockpiled together for easy disposal
- B Yes, this will help to avoid cross-contamination of the soil
- C Yes, this will help to make it easier to dispose of the soil
- D No, this is not a requirement under COSHH regulations

17.25 What should be the minimum capacity of a spillage bund around a fuel storage tank?

- A 130% of the total potential stored contents
- B 150% of the total potential stored contents
- C 110% of the total potential stored contents
- D 175% of the total potential stored contents

17.26 The programme of works you are managing has run behind and you will need the site to be operational on a Sunday. The operations will be noisy and there are nearby residents. What action should you take before you start the operations?

- A Inform the residents of your planned operation and offer hearing protection
- B Apply to the Local Authority for a Section 61 consent notice
- C Inform the residents of your planned operation and associated noise levels
- D Apply to the Local Authority for a Section 60 notice

17.27 To prevent pollution to watercourses, stockpiles and soil heaps must be placed away from drains and watercourses. What would be the most reasonably practicable solution to avoid silt run-off?

A Pumping the water run-off directly onto greenfield land

B Channelling the water run-off directly into the foul water sewer

C Channelling the water run-off directly onto brownfield land

D Installing geotextile silt fencing around the stockpiles and heaps

17.28 Is it acceptable to allow effluent that is produced from a concrete and cement washout to flow into a drain, watercourse or the ground?

A Yes, the washout is neutral, does not cause pollution, and does little harm to the environment

B No, the washout is highly alkaline, causes severe pollution, and is harmful to the environment

C No, the washout is highly acidic, causes severe pollution, and is harmful to the environment

D Yes, the washout is highly alkaline, does not cause pollution, and does little harm to the environment

17.29 Which **two** of the following would be the most effective way to prepare the workforce to efficiently respond to a pollution incident?

A Training in the correct use of spill kits

B Having an incident response plan displayed on site

C Using toolbox talks to discuss the incident response plan

D Undertaking simulated incident response exercises

E Having a sufficient number of spill kits around the site

17.30 You have identified a noise nuisance on site. Which **two** of the following would not be considered as 'best practical means' to control the nuisance?

A Ensuring that the workforce is competent and experienced

B Ensuring that the workforce's technical knowledge is current

C Consideration of the financial implications

D Consulting with the local community

E Consideration of the design, construction and maintenance of buildings

17.31 To avoid complaints, which one of the following would be the most effective way of managing dusty materials on site?

A Dampen them every day

B Store them under a cover

C Place them away from the site

D Provide face masks

17.32 Ground works are due to start around the periphery of an extensive site boundary and dust has been identified as a hazard. How can this work be controlled so air pollution risks are mitigated, therefore reducing the risks of statutory nuisance to the local community?

A Dampen down the site at the end of each dust-producing activity

B Conduct an assessment and apply benchmarked guidelines to test the deposits

C Complete daily visual assessments and stop work if dust levels rise

D Implement additional safety inspections to demonstrate active monitoring

17.33 During a recent site inspection, you have identified that waste is becoming a concern and you have decided to address this by delivering toolbox talks to the workforce. Which of the following statements are accurate?

1	Waste costs money to produce, and money to dispose of and remove	A	1 and 4 only
2	Rising landfill tax rates encourages the reduction of waste consigned to landfill	B	2 and 3 only
3	The true cost of disposing waste is around twenty times the cost of hiring a skip	C	3 and 4 only
4	The production of waste can be totally eliminated during the construction phase	D	1 and 2 only

17.34 You are managing a housing project and the work has created large amounts of surplus soil. Applying the waste hierarchy, which one of the following would be the most favoured option?

A Recycle the soil and distribute to other local projects, which promotes ecological transfer.

B Dispose of the soil to a local landfill site, which will help to balance soil pH levels.

C Reuse the soil for landscaping on the site, which will have minimal environmental impact.

D Recover the soil so that it can be used for composting, which will support biodiversity.

17.35 When disposing of construction and demolition waste, which one of the following is not a requirement?

A Contractors who carry or collect waste must have a waste carrier's licence

B Waste must be passed on to an authorised person

C Waste disposal facilities must have an incinerator

D All waste transfers must be supported by the correct document

17.36 Can waste electrical and electronic equipment (WEEE) be disposed of with other wastes?

A Yes, but only if it is put in scrap metal skips

B No, it must be disposed of separately

C No, it needs to be incinerated separately

D Yes, but only if it is mixed with other hazardous waste

17.37 You need to dispose of some plasterboard off-cuts. Can these go in with the general waste?

A Yes, it is classed as non-hazardous waste and can be mixed with other waste

B Yes, it is classed as hazardous waste and the off-cuts are insignificant in size

C No, it is classed as hazardous waste and should not be mixed with other wastes

D No, it is classed as non-hazardous waste and should not be mixed with other wastes

17.38 An employee decides to deliberately discharge the contents of a concrete and cement washout into a watercourse. Which one of the following could be held directly responsible for the pollution and be prosecuted?

A The site manager

B The company

C The employee

D The client

17.39 Which one of the following should be the primary consideration during all phases of a construction project?

A Waste recycling

B Waste disposal

C Waste hierarchy

D Waste storage

17.40 Which one of the following is not a class of waste in relation to the disposal to landfill?

A Inert waste

B Non-hazardous waste

C Hazardous waste

D Toxic waste

17.41 If a waste management plan is implemented on a construction site, which **two** of the following would be the main benefits?

A To ensure all forms of waste are recycled

B To help meet environmental obligations

C To ensure repeat business

D To maintain a competitive edge

E To help identify and minimise waste at the design stage

Specialist activities

The following specialist activities are included within the managers and professionals test and **all** need to be revised.

18 Demolition

18.01 If there are any doubts about the stability of a structure that is due for demolition, which **two** of the following would be deemed competent to provide assistance?

- A A civil engineer
- B A structural engineer
- C The company site manager
- D An environmental manager
- E A demolition engineer

18.02 The Construction (Design and Management) Regulations 2015 state that contractors must check that their workers are competent. Which two of the following are not deemed areas of competence?

1	Skills	A	1 and 2 only
2	Health	B	2 and 4 only
3	Experience	C	3 and 4 only
4	Fitness	D	1 and 3 only

18.03 What is the main reason that an inexperienced client should seek the assistance of a demolition engineer or consultant?

- A To ascertain the type and quantity of controlled explosives needed
- B To receive appropriate guidance on the appointment of duty holders
- C To formally apply for a Section 80 notice to permit the demolition work
- D To source utility companies' drawings that identify underground services

18.04 You are about to start a demolition project, and you are looking to employ a suitably-qualified supervisor. How would you know if they met the requirements of the National Demolition Training Group? They would hold...

A a Black card and a National Vocational Qualification (NVQ) Level 6 or higher

B a Green card and a National Vocational Qualification (NVQ) Level 3 or higher

C a Grey card and a National Vocational Qualification (NVQ) Level 3 or higher

D a Blue card and a National Vocational Qualification (NVQ) Level 2 or higher

18.05 When working in the demolition sector, an individual would be deemed a competent person if they...

A hold a National Demolition Training Group yellow card, a first-aid qualification and have experience of working on demolition sites

B have practical and theoretical knowledge, with actual experience of the type of demolition that is taking place on the site

C have experience in health and safety, and theoretical knowledge of the different types and methods of demolition work

D hold a National Demolition Training Group purple card, and have asbestos awareness and experience of working on demolition sites

18.06 In appointing a demolition contractor which one of the following would be the preferred method to gain information on their competence?

A Speak to their last client about the type of work, size of the project, whether there were any problems during the works they carried out, their performance and whether the client would use them again

B Speak to their last client about the contractor's fees, size of the project, whether there were any problems during the works, punctuality and whether the client would use them again

C Speak to their last client about the contractor's fees, size of the project, whether there were any problems during the works, punctuality and whether the client would recommend them

D Speak to their last client about the type of work, size of the project, whether there were any problems during the works, punctuality and whether the client would recommend them

18.07 Which one of the following factors should be considered when assessing an individual's competence before allowing them to work on a demolition project?

A Literacy levels

B Fear of heights

C Reaction to instructions

D Communication methods

18.08 You are managing the partial demolition of an office block built in 2010, which has been extensively damaged by recent storms. Before you start, you need to review safety-critical information on how the original structure was designed and built. In which one of the following documents would you find this?

A The pre-tender health and safety plan

B The demolition management file

C The construction phase plan

D The health and safety file

18.09 You have been tasked with carrying out demolition work that is near to overhead cables. Who must be contacted before work starts?

A The Health and Safety Executive (HSE)

B The local fire and rescue service

C The relevant power supply company

D The local authority

18.10 The risk of lead exposure is considered to be greater in which one of the following situations?

A Working on a commercial premises roof, installing lead slates, lead flashing and lead fixings

B Working in post-1970 buildings, especially during repair, maintenance, refurbishment and demolition

C Working on an industrial premises roof, stripping lead sheeting, lead flashing and lead underlays

D Working in pre-1970 buildings, especially during repair, maintenance, refurbishment and demolition

18.11 A noise assessment on a demolition project has found that 'Noise levels are likely to exceed the upper exposure action value'. What is this upper value, and what action would you take to reduce risk?

- A. The upper exposure action value is 85 dB(A). Use toolbox talks to communicate this issue to the workers who will be at increased risk, and issue hearing protection.
- B. The upper exposure action value is 85 dB(A). Create hearing protection zones that are clearly indicated with signs, and make hearing protection mandatory within these zones.
- C. The upper exposure action value is 95 dB(A). Use toolbox talks to communicate this issue to the workers who will be at increased risk, and issue them with hearing protection.
- D. The upper exposure action value is 95 dB(A). Create hearing protection zones that are clearly indicated with signs, and make hearing protection mandatory within these zones.

18.12 You are planning the demolition of a building, which involves identifying the associated hazards and risks. Which of the following would you consider as health hazards to the workforce?

1	Proximity concerns (tunnels, watercourses, etc.)	A	1 and 4 only
2	The type of construction and the materials used to build it	B	1 and 3 only
3	The age of the building and what it was previously used for	C	2 and 3 only
4	Separation points and their perceived load paths	D	2 and 4 only

18.13 Your company is demolishing a pre-1970s building. Workers may be at risk from exposure to lead dust, fumes and vapour from which of the following processes?

1	Removal of old doorframes and window frames by hand	A	1 and 2 only
2	Ripping out existing stud partition walls and ceiling boards	B	2 and 3 only
3	Stripping out and removing old gas pipework and guttering	C	2 and 4 only
4	Cutting of steel and metal materials using hot work processes	D	3 and 4 only

18.14 Which one of the following is considered to be a greater risk to health when working inside pre-1970 buildings, especially during repair, maintenance, refurbishment and demolition?

A Exposure to lead paint through dust and fumes

B Exposure to silica dust from mobile crusher operations

C Exposure to isocyanates from chemical cleaning products

D Exposure to Legionella bacteria from old stagnant water tanks

18.15 Which one of the following would be the primary health concern when demolition work is planned on a building that was constructed prior to 1999?

A The presence of oil

B The presence of lead

C The presence of fibreglass

D The presence of asbestos

18.16 What is the area called that receives materials from soft stripping or demolition operations?

A The dumping zone

B The landing zone

C The placing zone

D The waste zone

18.17 What does the acronym FOPS stand for?

A Falling object protective scaffold

B Flying object protective structures

C Falling object protective structures

D Flying object protective system

18.18 Which one of the following must be adhered to when planning demolition operations that require the use of plant and equipment?

A Construction (Design and Management) Regulations 2015 (CDM)

B Code of Practice for Full and Partial Demolition (BS 6187:2011)

C Code of Practice for Temporary Works Procedures (BS 5975:2019)

D Provision and Use of Work Equipment Regulations 1998 (PUWER)

18.19 You are risk assessing demolition work on a site where lifting equipment and excavators with high reach will be operating. Directly above, there are overhead power lines carrying 400kV on metal pylons. These lines must be isolated and disconnected, with written confirmation received on site before any demolition commences. Is this correct?

A. Yes - disconnection confirmation must be received in good time, and be acknowledged in writing by the principal contractor or contractor in control of the works

B. No - if isolation or disconnection are not possible, adequate control measures should be put in place following advice from the relevant power supply company

C. Yes - prior to work starting, disconnection confirmation in writing should be incorporated within both the construction phase plan and safe systems of work

D. No - if isolation or disconnection are not possible, the relevant power supply company will monitor demolition works on the site

18.20 You are working on a demolition project, and have been tasked with supervising the sorting of large quantities of different types of metal waste into segregated skips. You can use a mechanical excavator for this task, and must now select the most suitable attachment for this job, which is...

A. a hydraulic processor - as you will need to cut, crush and manipulate the waste materials

B. a grapple - as you are only handling and sorting light metals and general waste materials

C. a hyper-mobile arm - as you can lift very heavy material around by up to 360°

D. a bucket - as you can scoop up large quantities of waste with each operation

18.21 When using oxy-propane cutting equipment for a demolition task, which **two** of the following conditions must be met?

A. Cylinders are secured in a horizontal position, at least 5 m from any party wall, boundary fence line or hoarding

B. Cylinders are secured in an upright position, at least 3 m from any party wall, boundary fence line or hoarding

C. Hoses are secured with jubilee clips, not crimped fittings

D. Petrol or gas lighter are used to ignite the cutting torch

E. Flashback arresters are always fitted between cylinder gauges and hoses

18.22 When planning to use a super-high reach 360° excavator, which of the following factors relating to the ground conditions must be implemented to ensure safe use?

1	The excavator must be fitted with variable width tracks as it is heavier than a standard excavator	A	1 and 2 only
2	The excavator should not be used for crane or piling frame-level tasks due to its lower stability	B	2 and 3 only
3	The excavator needs to have a decreased ground-bearing pressure and reduced working envelope	C	1 and 4 only
4	The excavator should be segregated from areas where any voids or ducts have been identified	D	3 and 4 only

18.23 You are reviewing a risk assessment with the operator of a high-reach machine before they start work on part of a demolition project. Why is it important that they are made aware that the base of the high-reach machine does not sit within the red zone?

A The operator will have a restricted view of the structure that they are working on

B The operator is unlikely to have adequate protection if large pieces of the structure fall down the boom and hit the cab

C The operator will be at a greater risk from dust, silica, and asbestos hazards

D The operator must follow strict procedures that prohibit the boom from being used at an angle greater than 50°

18.24 The level of supervision required on potentially hazardous demolition work is dependent on which of the following?

1	The planning and management abilities of all involved	A	1 and 2 only
2	The individuals' education, physical agility, literacy and attitude	B	1 and 4 only
3	The training, knowledge, experience and skills of the workforce	C	2 and 3 only
4	The communication of control measures and precautions	D	3 and 4 only

18.25 When developing a construction phase plan that incorporates demolition work, which one of the following would not need to be included?

A Contractor interviews and tender documents

B Project information and scope of work

C Existing environmental information and drawings

D Risk assessment and special hazards

18.26 Which is the preferred method of demolishing brick or internal walls by hand?

A Undercut the wall at ground level

B Work in the reverse order to construction

C Work in the same order as construction

D Remove at corners and collapse in sections

18.27 Which one of the following does not support good health and safety standards on a demolition project?

A Ensuring that safe systems of work are in place and adhered to

B Appointing a suitably-competent demolition contractor

C Informing site workers of the location of the asbestos survey

D Ensuring that there is sufficient time to gain licences and permits

18.28 When operating a mobile crusher, and processing materials created by a demolition project for reuse on site, which one of the following procedures should be avoided?

A Manually removing debris when the crusher has stopped and is isolated

B Checking the machinery guarding on the crusher at frequent intervals

C Standing on the access platform while the crusher is running

D Operating the crusher at the same time as other plant

18.29 In which one of the following common demolition processes would a 360° adapted demolition rig with multi-functional attachments be used?

A For situations where the machine operator is isolated and does not have visibility of all of the structure's elevations

B For relatively low-level buildings or after other height reduction techniques have been carried out

C For undermining or undercutting of structures when a machine is unable to reach the top of the building

D For poor ground conditions where there is increased ground-bearing pressure and extended working envelopes

18.30 Your company has been appointed as principal contractor for the decommissioning and demolition of a disused telecommunications centre. For this type of project, which **two** of the following are good practice recommendations identified in the code of practice for full and partial demolition (BS 6187:2011)?

A. Appointment of a nationally-registered demolition contractor

B. Acquiring a knowledge of the site, including its former uses

C. Identifying opportunities for materials to be reused, reclaimed and recycled

D. Identifying and establishing responsibilities during the demolition process

E. Giving notice to the Local Authority Building Control department

18.31 The height of a wall or building to be demolished should not normally be greater than the maximum reach of the machine being used to carry out the work. Is this correct?

A. No - the machine operator takes into account the height of the structure being demolished and the buildup of debris

B. Yes - this ensures that a safe distance is being maintained between the machine and the structure being demolished

C. No - the machine operator can construct a ramp, using rubble from previous demolition for the machine to sit on to increase the reach

D. Yes - this ensures that the machine operator is able to clearly see all elevations of the structure being demolished

18.32 Under the Certification of Competence of Demolition Operatives (CCDO) card scheme, what colour card would a demolition labourer have?

A. White

B. Yellow

C. Red

D. Green

18.33 The demolition industry has developed a process called DRIDS, which explains how to identify, reuse or recycle waste. What does the abbreviation DRIDS stand for?

A. Demolition and recycling information datasheets

B. Demolition and refurbishment information datasheets

C. Demolition and reuse information datasheets

D. Demolition and renewal information datasheets

19 Highway works

19.01 When normal traffic management arrangements are not feasible due to restricted road width, which method of work should be used, and what key safety measures do they provide?

A Safety zone working - traffic speeds must be reduced to 20 mph or less alongside the working space

B Convoy working - traffic is directed through the works by an appropriately-signed works vehicle

C Chicane working - traffic must be released in small quantities by the use of Stop/Go boards

D Shuttle working - traffic signals are used to alternate flows on a one-way section of the road

19.02 Where trench excavations have to be left open as part of ongoing water mains replacement works in a pedestrianised area, they must be covered with road plates at night in order to open the footway. Is this correct?

A Yes - excavations in pedestrianised areas should be covered with road plates at night, and their use should be planned in advance to identify the appropriate size and thickness

B No - excavations should be adequately guarded at night, and where necessary should be fitted with lights where construction workers or the public could be put at risk of falling

C Yes - excavations should be covered with road plates at night, and the surface of the plates should be non-slip, as well as free of trip hazards, raised edges or overlaps on boarding

D No - excavations in pedestrianised areas should be adequately guarded at night, and fitted with temporary covers if required, taking the needs of the general public into account

19.03 How must highways signage and guarding equipment be properly secured?

A With kerbstones

B With ground fixtures

C With built-in weights

D With chains

19.04 If there is the potential for mud from a site to create a hazard on public roads, what should be the site manager's primary action?

A Create a dedicated team to clean the road regularly

B Employ an effective on-site wheel-washing system

C Employ a mechanical road sweeper to clean the road

D Erect road signs to warn road users of the mud hazard

19.05 In which of the following documents would you find information about the distances for setting out highways signs?

1 The Traffic Signs Manual

2 The Code of Practice (Green Book)

3 The Safe System of Work

4 The Code of Practice (Red Book)

A 1 and 2 only

B 2 and 3 only

C 3 and 4 only

D 1 and 4 only

19.06 On a single carriageway that has a speed limit of 30 mph or less, what are the minimum requirements for the visibility of the first sign for roadworks, in terms of placement and size?

A At least 40 m distance, and 650 mm in size

B At least 50 m distance, and 700 mm in size

C At least 60 m distance, and 600 mm in size

D At least 30 m distance, and 500 mm in size

19.07 Which one of the following is not a requirement for inspecting highways signs on an attended site?

A Every time you start work

B Regularly during active work

C After every three hours

D Before you leave the site

19.08 You are managing a highways works team, and the weather forecast indicates that heavy rain and mist are moving towards your location. Which one of the following should be your primary action?

A Brief the team on the situation, ensure that each worker has wet weather clothing, and provide them with strobe lights to make them more visible in the poor conditions

B Brief the team on the situation, deploy additional signs to alert road users of the incoming weather, and then abandon operations immediately

C Brief the team on the situation, ensure that additional signs are available and ready to be deployed, and monitor the weather with a view to abandoning operations if conditions become poor

D Brief the team on the situation, ensure that additional signs and strobe lights have been deployed, and prepare the workers to abandon operations when conditions start to become poor

19.09 For mobile works on a dual carriageway with a speed limit of 40 mph or less, which one of the following safety measures is recommended?

A Placing appropriately-sized signs in the central reservation

B Safeguarding operatives with an impact protection vehicle

C Providing and displaying various vehicle-mounted signs

D Not allowing frequent work stops of more than 15 minutes

19.12 What action is required when a highways vehicle fitted with a direction arrow is travelling from site to site?

A Point the direction arrow upwards

B Make the visibility of the direction arrow clear

C Point the direction arrow downwards

D Cover or remove the direction arrow

19.10 When conducting work on a single carriageway, works vehicles can be parked in front of the work area to provide some physical protection to the driver. Which two of the following are the correct minimum distances, dependent upon the imposed road speed, that must be maintained between the vehicle and work area?

1	Two metres, for a speed limit of 30 mph or less	A	1 and 2 only
2	Five metres, for a speed limit of 40 mph or more	B	2 and 3 only
3	Eight metres, for a speed limit of 30 mph or less	C	1 and 4 only
4	Ten metres, for a speed limit of 40 mph or more	D	3 and 4 only

19.11 Which **two** of the following situations require a highways team operative to switch on the amber flashing beacon that is fitted to a highways vehicle?

A At all times when the vehicle is being used

B At all times during short-duration static works

C When travelling to and from the site compound

D In poor weather conditions if lighting levels or visibility is poor

E When the vehicle is travelling slower than the speed of traffic

19.13 When planning highway works that require setting up portable traffic signals to ensure the safe movement of vehicles, which one of the following statements applies?

A The relevant highways authority must be notified of works requiring the use of portable traffic signals 14 days prior to starting the works

B The relevant highways authority must be notified of the requirement for portable traffic signals no later than 10:00 on the morning of the works

C The relevant highways authority and local council must give written authorisation of works requiring the use of portable traffic signals

D The relevant highways authority must be informed so that authorisation can be obtained for the works to use portable traffic signals

19.14 When conducting work underneath or in close proximity to overhead power lines, it should not be done during darkness or poor visibility weather conditions. Is this correct?

A Yes, as the operator's vision will become significantly impaired and could confuse the distance of the power lines

B No, as there are specific safe systems and monitoring procedures in place for operating around power lines

C Yes, as the reflection from the vehicles lighting can obscure rather than illuminate the power lines

D No, as the vehicles are fitted with specialist external lighting that clearly illuminates the power lines

19.15 Which **two** of the following vehicle requirements must be met when carrying out short-duration static works on a dual carriageway?

A The vehicle must have one or more amber warning beacons, which should only be activated if travelling within the works area

B The vehicle must have one or more amber warning beacons remaining on at all times, so that at least one beacon can be seen from any direction

C Vehicle-mounted 'Keep right/left' signs must be clearly displayed when the vehicle is travelling to and from the works area

D A 'Keep right/left' sign must be displayed for drivers approaching the works on the same side of the carriageway, showing which side to pass

E The vehicle must have two or more amber warning beacons, so that at least two beacons can be seen by drivers approaching the works

19.16 Which **two** of the following must apply when planning to control traffic movements at highway works using the 'give and take' method of traffic control?

A Drivers approaching the works are made aware that the speed limit is 40 mph or less

B No more than 25 heavy goods vehicles or buses can pass the works every hour

C Drivers approaching from either direction can see 50 metres beyond the end of the works

D The length of the works from the first cone to the last cone is 75 metres or less

E Two-way traffic flow is no more than 20 vehicles counted over 3 minutes passing the works

19.17 For operatives cutting grass verges near to a B-class road, which of the following is an example of the minimum acceptable level of high-visibility protection?

A High-visibility trousers with two 5 cm reflective bands around each leg

B High-visibility vest with two 5 cm bands of reflective tape around the body

C High-visibility vest with three-quarter length sleeves and two 5 cm bands of reflective tape around the body

D High-visibility jacket and trouser suit with two 5 cm bands of reflective tape around the body and arms

19.18 A high-visibility vest or jacket does not need to be worn by operatives undertaking resurfacing operations on a road that is not maintained at public expense.
Is this correct?

A No - the minimum level of protection required for anyone working on a private road is a high-visibility vest with two 5 cm bands of reflective tape around the body

B Yes - the minimum level of protection required for anyone working on a private road is high-visibility trousers with two 5 cm reflective bands around each leg

C No - the minimum level of protection required for anyone working on a private road is a high-visibility vest with two 5 cm bands of reflective tape around the body, and braces over both shoulders

D Yes - the minimum level of protection required for anyone working on a private road is high-visibility overalls with one 5 cm band around the body, and two 5 cm reflective bands around each leg

19.19 What is the main reason that personal protective clothing worn during road work operations should always be kept in a clean and usable condition?

A It is a legal requirement for all employees

B Dirt or damage will reduce its protection against rain

C Dirt build up could result in a health condition

D Dirt or damage will reduce its high-visibility

19.20 When carrying out street or road works operations, which one of the following items of personal protective equipment should always be worn?

A Safety glasses

B Hearing defenders

C Safety footwear

D Dust mask

19.21 Roadworks operations are taking place next to a busy dual carriageway. Which **two** priorities for hearing protection need to be considered?

A The hazard of noise takes priority over the hazard of traffic

B A suitable risk assessment is conducted

C Protection is only used at peak traffic intervals

D Protection is sufficiently maintained and stored

E The hazard of traffic takes priority over the hazard of noise

19.22 You are planning a road works operation, and the risk assessment indicates that Class 3 clothing will be required. What type of environment would you be operating in?

A On a B-class road, which requires clothing with intermediary visibility

B On a motorway, which requires clothing with the highest visibility

C On a private road, which requires clothing with the lowest visibility

D On a dual carriageway, which requires clothing with intermediary visibility

19.23 You are about to start road works operations on a B class road. Which one of the following documents must determine the appropriate class of high-visibility clothing to be worn for this work?

A The method statement

B The permit to work

C The risk assessment

D The safe system of work

19.24 What is the main reason for wearing appropriate high-visibility clothing when conducting street or roadworks operations?

A To reflect harmful UV rays that could cause health conditions

B To keep you warm during the autumn and winter months

C To keep you cool during the spring and summer months

D To be clearly seen by colleagues and the travelling public

Further information

01 General responsibilities

1.01	C	1.29	D
1.02	C	1.30	B
1.03	A, E	1.31	C
1.04	D, E	1.32	C
1.05	D	1.33	D
1.06	A	1.34	A
1.07	A	1.35	A, C
1.08	C	1.36	B
1.09	A, D	1.37	C, E
1.10	C	1.38	B
1.11	C	1.39	C
1.12	B	1.40	A, C
1.13	B	1.41	C
1.14	A	1.42	D
1.15	A	1.43	B
1.16	A, C	1.44	C
1.17	B, E	1.45	C
1.18	D	1.46	B, C
1.19	D	1.47	D
1.20	A	1.48	B
1.21	A	1.49	A
1.22	C	1.50	B
1.23	B	1.51	A
1.24	C, E	1.52	C, E
1.25	B	1.53	C
1.26	C	1.54	B
1.27	B	1.55	B
1.28	B	1.56	B

1.57	C	1.62	C
1.58	C	1.63	A
1.59	A	1.64	A
1.60	C	1.65	B
1.61	A, C		

02 Accident prevention

2.01	C	2.16	A, D
2.02	C	2.17	B
2.03	A, D	2.18	A
2.04	B	2.19	A
2.05	B	2.20	C
2.06	B, C	2.21	B, D
2.07	A	2.22	D
2.08	D	2.23	C
2.09	D	2.24	A
2.10	B	2.25	C
2.11	C	2.26	A
2.12	C	2.27	D
2.13	A	2.28	D
2.14	B	2.29	B
2.15	A	2.30	D

03 Construction (Design and Management) Regulations 2015 (CDM)

3.01	A	3.05	C
3.02	A	3.06	A
3.03	D	3.07	A
3.04	C	3.08	B

3.09	D	3.17	C
3.10	B	3.18	B
3.11	D	3.19	D, E
3.12	A	3.20	C
3.13	B, D	3.21	C
3.14	B	3.22	A, C, D
3.15	C	3.23	B, C
3.16	C	3.24	B

04 Occupational health and welfare

4.01	A	4.20	C
4.02	A	4.21	A
4.03	C	4.22	A
4.04	A	4.23	B
4.05	B	4.24	B
4.06	A, E	4.25	C
4.07	C	4.26	B
4.08	C	4.27	A, E
4.09	B	4.28	A
4.10	A	4.29	A
4.11	A	4.30	B
4.12	B	4.31	C, E
4.13	C	4.32	C
4.14	B, C	4.33	D
4.15	B	4.34	A
4.16	D	4.35	A
4.17	A	4.36	D
4.18	C	4.37	B
4.19	A	4.38	D

4.39	A, C	4.48	D
4.40	B	4.49	D
4.41	C	4.50	C
4.42	C	4.51	A
4.43	B	4.52	A
4.44	A	4.53	D
4.45	D	4.54	A, C
4.46	A, C	4.55	A
4.47	C	4.56	D

05 First aid and emergency procedures

5.01	B	5.19	C
5.02	B	5.20	C
5.03	C	5.21	B, E
5.04	D	5.22	D
5.05	A, B, F	5.23	B
5.06	C	5.24	D
5.07	B	5.25	C
5.08	D	5.26	B
5.09	D	5.27	B, C, F
5.10	D	5.28	C
5.11	D	5.29	D
5.12	B	5.30	C
5.13	B, C, F	5.31	D
5.14	C	5.32	C
5.15	C	5.33	B
5.16	C	5.34	C
5.17	C	5.35	C
5.18	C	5.36	D

5.37	B	5.39	C
5.38	B	5.40	A, D, F

06 Personal protective equipment

6.01	A	6.26	A, C
6.02	A	6.27	D
6.03	A	6.28	D
6.04	A	6.29	C
6.05	A	6.30	C
6.06	B	6.31	C
6.07	A	6.32	B
6.08	A, C	6.33	B
6.09	A	6.34	D
6.10	B	6.35	A
6.11	A	6.36	D
6.12	A	6.37	C
6.13	A	6.38	A, D
6.14	A	6.39	D
6.15	B	6.40	B
6.16	A, C	6.41	B
6.17	B, E	6.42	C
6.18	A	6.43	A
6.19	A	6.44	D
6.20	A	6.45	B, C
6.21	B	6.46	C
6.22	A	6.47	D
6.23	A, D	6.48	C
6.24	A, D	6.49	D
6.25	A	6.50	B

6.51	B
6.52	B
6.53	A, C
6.54	C
6.55	D
6.56	B
6.57	A
6.58	A, D
6.59	D
6.60	C
6.61	B
6.62	A, D
6.63	B
6.64	A
6.65	A
6.66	A

07 Dust and fumes (Respiratory hazards)

7.01	D
7.02	A
7.03	A
7.04	A, B
7.05	A, B
7.06	A
7.07	A
7.08	B
7.09	A, D
7.10	D
7.11	D
7.12	D
7.13	D
7.14	A
7.15	D
7.16	B
7.17	C
7.18	B
7.19	A
7.20	A
7.21	B
7.22	B
7.23	A, C
7.24	B, D

08 Noise and vibration

8.01	A
8.02	A
8.03	A, D
8.04	C
8.05	D
8.06	A
8.07	C
8.08	D
8.09	A
8.10	D

8.11	A
8.12	C
8.13	B
8.14	D
8.15	D
8.16	C, E
8.17	B
8.18	B
8.19	D
8.20	A
8.21	D
8.22	D
8.23	C
8.24	A
8.25	B
8.26	A, C
8.27	C
8.28	C
8.29	A
8.30	C
8.31	A
8.32	A
8.33	D
8.34	B
8.35	A
8.36	A, B
8.37	A
8.38	B
8.39	C
8.40	B

09 Hazardous substances

9.01	A
9.02	D
9.03	A, D, E
9.04	C
9.05	B
9.06	B
9.07	A
9.08	C
9.09	D
9.10	B
9.11	C
9.12	A
9.13	A
9.14	A
9.15	D
9.16	C, D
9.17	B
9.18	A
9.19	B
9.20	A, C, D
9.21	C
9.22	A
9.23	D
9.24	D

9.25	C
9.26	D
9.27	D
9.28	A
9.29	B
9.30	B
9.31	A, C, E
9.32	D
9.33	A
9.34	D
9.35	A
9.36	A
9.37	C
9.38	B
9.39	D
9.40	A
9.41	B, D
9.42	D
9.43	B
9.44	C
9.45	A
9.46	A
9.47	D
9.48	B
9.49	A, C, E
9.50	C
9.51	B
9.52	D
9.53	A
9.54	C
9.55	B, D
9.56	C
9.57	A
9.58	D
9.59	B
9.60	B
9.61	C
9.62	D
9.63	B
9.64	D
9.65	A
9.66	B
9.67	B
9.68	C
9.69	A
9.70	B
9.71	A, B, D
9.72	C

10 Manual handling

10.01	A
10.02	A
10.03	A
10.04	A, B, D, F
10.05	A
10.06	A, C, E

10.07	D
10.08	A, C, D, F
10.09	D
10.10	B
10.11	C
10.12	D
10.13	C
10.14	B, C, D
10.15	A
10.16	A, D, E
10.17	C, E
10.18	B, C, E
10.19	A, B, D
10.20	B
10.21	B
10.22	A
10.23	D
10.24	B
10.25	B, C
10.26	B
10.27	B
10.28	E, F
10.29	A
10.30	C
10.31	B
10.32	A
10.33	A
10.34	B
10.35	B
10.36	B, D
10.37	C
10.38	D, E
10.39	A

11 Safety signage and symbols

11.01	C
11.02	D
11.03	B
11.04	A
11.05	A
11.06	C
11.07	B
11.08	D

12 Fire prevention and control

12.01	D, E
12.02	A
12.03	D
12.04	B
12.05	C
12.06	A
12.07	A
12.08	D

12.09	B
12.10	C
12.11	A, C, E
12.12	D
12.13	C
12.14	D
12.15	D
12.16	C
12.17	C
12.18	D
12.19	D
12.20	C
12.21	A
12.22	B, C, D
12.23	D
12.24	A
12.25	B
12.26	D
12.27	B
12.28	A, C, E
12.29	C
12.30	B
12.31	C
12.32	B
12.33	B
12.34	C
12.35	D
12.36	C
12.37	A
12.38	D, E
12.39	A
12.40	A
12.41	D
12.42	B
12.43	A
12.44	C
12.45	D
12.46	C
12.47	B, C, E
12.48	B

13 Electrical safety, tools, equipment, lasers and drones

13.01	A
13.02	B
13.03	A
13.04	C
13.05	A
13.06	D
13.07	A, D, E
13.08	B
13.09	A, C, E
13.10	D
13.11	C
13.12	D
13.13	B
13.14	A

13.15	B	13.24	A
13.16	A	13.25	A
13.17	C	13.26	A
13.18	C	13.27	B
13.19	B, C, D	13.28	C
13.20	A	13.29	C, D, E
13.21	B	13.30	B
13.22	D	13.31	A
13.23	A, C, E	13.32	C

14 Site transport safety and lifting operations

14.01	C, E	14.19	C
14.02	B	14.20	D
14.03	D	14.21	B
14.04	A	14.22	D
14.05	C	14.23	A
14.06	A, C	14.24	D
14.07	A	14.25	D
14.08	C	14.26	C
14.09	B, D	14.27	C
14.10	B	14.28	B
14.11	B	14.29	B, C
14.12	C	14.30	C
14.13	B	14.31	C
14.14	C	14.32	B
14.15	A, B, E	14.33	A
14.16	B	14.34	B
14.17	B, C	14.35	D
14.18	C	14.36	C, E

14.37	C, E
14.38	A
14.39	D
14.40	B, E
14.41	A
14.42	B
14.43	A, C
14.44	B, C
14.45	B
14.46	B
14.47	A, C
14.48	C
14.49	D
14.50	D
14.51	B
14.52	B
14.53	D
14.54	A
14.55	D

15 Working at height

15.01	A
15.02	A, C, E
15.03	B
15.04	D
15.05	C
15.06	B
15.07	B
15.08	C
15.09	D
15.10	A
15.11	A, D
15.12	D
15.13	B
15.14	B
15.15	D
15.16	C
15.17	B
15.18	C
15.19	C
15.20	D
15.21	A
15.22	B
15.23	A, C, E
15.24	A
15.25	C
15.26	C
15.27	A, B, D
15.28	A
15.29	A
15.30	C
15.31	C
15.32	C
15.33	B
15.34	C

15.35	D
15.36	C
15.37	A
15.38	B
15.39	C
15.40	D
15.41	C
15.42	B
15.43	A
15.44	B
15.45	D
15.46	C
15.47	B
15.48	B
15.49	C
15.50	D
15.51	A
15.52	A, C, D
15.53	D
15.54	B
15.55	B
15.56	B
15.57	C
15.58	B
15.59	B, C, D
15.60	A
15.61	C
15.62	D
15.63	D
15.64	D
15.65	B, C, D
15.66	A
15.67	A
15.68	B
15.69	C
15.70	B
15.71	D
15.72	A
15.73	B, C, E
15.74	B
15.75	A
15.76	A
15.77	D
15.78	D
15.79	D
15.80	C
15.81	B
15.82	D
15.83	C
15.84	D
15.85	A, C, D
15.86	A
15.87	B
15.88	B
15.89	B
15.90	C

16 Excavations and confined spaces

16.01	B	16.13	D
16.02	D	16.14	B
16.03	A	16.15	C
16.04	B	16.16	B
16.05	B	16.17	A, D, E
16.06	A	16.18	B, D
16.07	D	16.19	A, D
16.08	C, E	16.20	D
16.09	D	16.21	C
16.10	B	16.22	D
16.11	D	16.23	D
16.12	B	16.24	B

17 Environmental awareness and waste control

17.01	C	17.15	A, C
17.02	B, E	17.16	A
17.03	B	17.17	A, C
17.04	B	17.18	C
17.05	C	17.19	A
17.06	C	17.20	B
17.07	B	17.21	C
17.08	B	17.22	C
17.09	B	17.23	B
17.10	A	17.24	B
17.11	B	17.25	C
17.12	D	17.26	B
17.13	C	17.27	D
17.14	B	17.28	B

17.29	A, D
17.30	A, D
17.31	B
17.32	B
17.33	D
17.34	C
17.35	C
17.36	B
17.37	D
17.38	C
17.39	C
17.40	D
17.41	B, E

18 Demolition

18.01	B, E
18.02	B
18.03	B
18.04	B
18.05	B
18.06	A
18.07	A
18.08	D
18.09	C
18.10	D
18.11	B
18.12	C
18.13	D
18.14	A
18.15	D
18.16	B
18.17	C
18.18	D
18.19	B
18.20	A
18.21	B, E
18.22	C
18.23	B
18.24	C
18.25	A
18.26	B
18.27	C
18.28	C
18.29	B
18.30	B, D
18.31	B
18.32	D
18.33	B

19 Highway works

19.01	B	19.13	D
19.02	B	19.14	C
19.03	C	19.15	B, D
19.04	B	19.16	C, E
19.05	D	19.17	B
19.06	C	19.18	B
19.07	C	19.19	D
19.08	C	19.20	C
19.09	B	19.21	B, E
19.10	A	19.22	B
19.11	B, E	19.23	C
19.12	D	19.24	D